Music Technology in Live Performance

T0256301

Music Technology in Live Performance explores techniques to augment live musical performance and represents a comprehensive guide to best practices in music technology for live performance.

This book presents a practical and accessible introduction to the theories of liveness and an array of live performance technologies and techniques. Areas covered include analogue and digital audio, live sound, the recording studio, and electronic music, revealing best professional practices and expert tips, alongside an exploration of approaches to increasing the exchange of energy in live performance.

Music Technology in Live Performance is an ideal introduction for students of music performance, music production, and music technology, and a vital resource for professional musicians, producers, and technology developers.

Tim Canfer is a lecturer in music production at Keele University. He is a musician, a music producer, and a technology developer (including the www.reactivebacking.com suite of Max for Live plugins). Tim is the host and producer of the podcast *Sound Learnings* and a series editor for our new *Perspectives on Education in Audio & Music Production* series.

Music Technology in Live Performance

Tools, Techniques, and Interaction

TIM CANFER

Routledge
Taylor & Francis Group

LONDON AND NEW YORK

Designed cover image: © samserius – stock.adobe.com

First published 2024
by Routledge
4 Park Square, Milton Park, Abingdon, Oxon OX14 4RN

and by Routledge
605 Third Avenue, New York, NY 10158

Routledge is an imprint of the Taylor & Francis Group, an informa business

British Library Cataloguing-in-Publication Data
A catalogue record for this book is available from the British Library

Library of Congress Cataloging-in-Publication Data
Names: Canfer, Tim, author.
Title: Music technology in live performance : tools, techniques,
and interaction / Tim Canfer.
Identifiers: LCCN 2023027664 (print) | LCCN 2023027665 (ebook) |
ISBN 9781032440897 (paperback) | ISBN 9781032440910 (hardback) |
ISBN 9781003370406 (ebook)
Subjects: LCSH: Music–Performance. | Computer sound processing. |
Concerts–Production and direction. | Electronic music–Production and direction.
Classification: LCC ML457 .C37 2023 (print) | LCC ML457 (ebook) |
DDC 780.78–dc23/eng/20230912
LC record available at https://lccn.loc.gov/2023027664
LC ebook record available at https://lccn.loc.gov/2023027665

ISBN: 978-1-032-44091-0 (hbk)
ISBN: 978-1-032-44089-7 (pbk)
ISBN: 978-1-003-37040-6 (ebk)

DOI: 10.4324/9781003370406

Typeset in Sabon
by Newgen Publishing UK

Access the Support Material: www.routledge.com/9781032440897

CONTENTS

ACKNOWLEDGEMENTS

Thanks to the quartet-of-chaos and the trio-of-tolerance for their continued support and patience, and to everyone who has helped with this project – in particular Russ, Rich T, Susan, Bill, and Spag.

NOTE ABOUT FIGURES

All of the figures in this book are available as pdf files that can be found on this webpage www.routledge.com/9781032440897 or bit.ly/tc_mtlp or use the QR code shown below:

Diagrams are made in full colour for use in teaching; please attribute using an appropriate referencing style. Screenshots are also included for screen clarity.

Introduction:
Exchange of energy

In a 2016 interview, Johnny Marr,[1] guitarist for The Smiths, Modest Mouse, and many others – revealed his early aversion to playing live: "I didn't like playing gigs. For me, it was always about making that music." Later in his career, he learned how to front a band himself. He discovered that live performance is: "a great scenario. I don't mean the praise and the adulation; I mean what a gig is: the exchange of energy."

The phrase 'exchange of energy' is a fitting metaphor for what happens at a live musical performance event and neatly sums up the focus of this book.

Musicians have performed for audiences for centuries. Before the electrical stage of microphones and amplifiers, music came in direct acoustic form, clearly caused by musicians' actions, so a rich understanding of musical cause and effect has developed. The conventions for musicians to demonstrate their skilled performance and allow an exchange of energy have been established and refined.

Since the electrical stage and, more recently, the digital stage (these stages are explained in Section 2), modern technologies have enabled a vast range of exciting sonic possibilities. However, these technologies are predominantly used to create recordings. If the amount of recorded music that is listened to is compared to the amount of live music listened to in a live event, recorded music clearly far outweighs live music. This cultural dominance of recorded music makes it the benchmark that live music must often measure up to.

As the name suggests, the 'liveness' of a musical performance event describes how live that event is and exists in contrast to recordings, or what this book describes as 'non-live musical elements'. Liveness is fundamental to giving meaning to and establishing the appeal of live performance events, leading to the exchange of energy and the other factors that make live performance events so exciting and vital: the immediacy, uniqueness, and authenticity of the moment.

Using non-live musical elements (i.e., samples, backing tracks, sequencers) is likely to reduce the liveness of a live performance event. For example, a group playing exclusively acoustic instruments has significantly more liveness than another group playing to a fixed backing track. Along with this reduced liveness comes an interference with musical cause and effect and, accordingly, the exchange of energy in musical performance events. That there has been a cultural acceptance of this lack of liveness is understandable, given the historic technological limitations. Those with a long experience in electronic music performances

DOI: 10.4324/9781003370406-1

are likely to take this lack of liveness as integral to electronic music's performance aesthetic. However, developments in modern technologies and their associated techniques mean that this lack of liveness need not necessarily be as much of an issue in contemporary music performance events anymore.

This book explores innovative, experimental, and interactive live music performance tools and techniques that work towards a more effective exchange of energy (Chapter 5). The underpinning areas of live sound (Chapter 1), the recording studio (Chapter 2), electronic music (Chapter 3), and the tools of music technology in live performance (Chapter 4) are also covered, going from foundations to best practices.

The rudiments of liveness theory are introduced in this section and in Section 5.1.3 to give broader context and critical awareness to the techniques explored in Chapter 5. A more thorough academic exploration of the subject will be presented in my forthcoming publication about liveness in modern music performance.

I.1 EXCHANGE OF ENERGY UNPACKED

For anyone who has appreciated a live musical performance, the exchange of energy metaphor is likely to make intuitive sense. However, for the metaphor to be of practical use, its meaning needs explaining, expanding, and clarifying.

Three questions can effectively unpack 'exchange of energy':

- *What* is the energy?
- *How* is it exchanged?
- *Who* is exchanging it?

What? In this case, 'the energy' itself is expression. For musical performers, this is likely to be a combination of musical expression, expression of movement, and the narrative expression of words and lyrics. Typically, the energy is a combination of many different musical expressions performed together to form aesthetic meaning.

How? The form of exchange is in the sound and visuals of the performance.

Who? The different agents in this exchange are the performers, the audience members, and the systems used.

Considering these points gives us three different exchange of energy models: traditional, backing, and interactive. These models are explained in the following sections and are shown in Figures i.1, i.2, and i.3.

i.1.1 Exchange of energy – traditional

In the traditional model shown in Figure i.1, and as the core of the other two models, the performers-to-audience arrow shows a direct and significant exchange of musical expression, indicated by a filled and wide arrow.

The audience-to-performers arrow indicates less energy exchanged than performers-to-audience as the performance is listened to and observed. However, this arrow reflects that for a performance to have any significant exchange of energy, it helps if the performers react and connect to the audience.

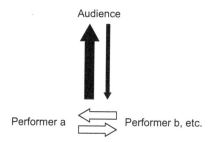

FIGURE I.1
Traditional exchange of energy model

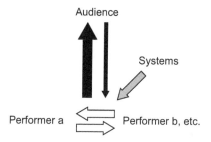

FIGURE I.2
Exchange of energy model with a backing track system

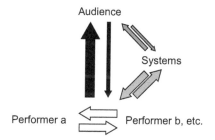

FIGURE I.3
Exchange of energy model with interactive systems

The energy exchanged between performers is less direct, indicated by the unfilled arrow. This is because the activity of playing will typically be well known to each musician, and when successful, the exchange is likely to be in the nuanced performance differences between each musician.

In the traditional exchange of energy model, all elements are live. There is no use of recordings or any pre-produced media other than the broader pre-production elements, such as a set list, the musical arrangements themselves, and the genre conventions (e.g., a 12-bar blues jam).

i.1.2 Exchange of energy – backing

The backing model in Figure i.2 shows a similar setup but with a different kind of one-way exchange of energy as the backing, or click track, can only lead the performance with the performers having to follow. A backing system is a machine form of expression rather than a human one, where the expression is in the form of a fixed audio recording playback system (or a fixed MIDI, or mixed audio and MIDI playback system). This machine-driven exchange of energy is shown as the grey filled arrows in Figures i.2 and i.3.

The backing track model is a generalised but common setup where several musical elements are live. The backing track here is fixed, so it is a non-live musical element.

i.1.3 Exchange of energy – interactive

The interactive model in Figure i.3 builds further on the backing model; however, the systems now allow machine-to-human interaction. Grey filled arrows indicate the machine-driven energy exchanged between systems-to-performers and audience-to-system. Unlike the backing model, this exchange is two-way. An interactive system has the potential both to react to performer data and inform the performance, and to react to audience data and inform how the performance is perceived.

As discussed in Figure i.1, the size of the arrows indicates the likely relative amounts of exchange. As the systems are typically integrated into the performance setup, performers-to-systems interactions are likely to be more significant than audience-to-systems interactions.

All three models discussed here are inherently general, covering a broad range of setups. Specific systems that allow these interactions between systems-to-performers are explored in Chapter 5. Specific systems-to-audience interactions are outside the scope of this book but include the following setup types: Reactive audience-to-system setups that are likely to be in the form of camera or audio-driven input and image processing and audio input that can detect different kinds of audience movement and sound. This data can then be mapped onto musical parameters. Other types of mapping are discussed in Section 5.1.5.

i.1.4 Fine expressive control vs sonic capabilities

Exchange of energy in musical performance is a by-product of instrument design development – specifically, the trade-off between the fine expressive control of an instrument and greater sonic capabilities.

We can start, for example, with the violin, invented in the 16th century. A violin gives an expert violinist a high degree of fine expressive control over the instrument's pitch and dynamics. The performer's tiniest movement fundamentally influences the music's character; however, the violinist can only maintain that level of control when playing one note at a time.

We can then compare the violin to a more mechanical instrument, such as the piano, invented 200 years later in the 18th century. The piano allows an expert pianist to play as many notes as they have keys. However, a pianist cannot vary the pitch or dynamics of the notes once played and can only play the notes tuned. The mechanics of the piano trade in the fine expressive monophonic control of the violin for a massive 88-note polyphony (notes that can be played at once).

Moving another 200 years into the 20th century, electronic instruments such as synthesisers and samplers have dramatically expanded on the polyphony of the piano, allowing an almost infinite combination of sounds and instrument layers. Much like the mechanisms of the piano that determine the character of the piano's tone, these internal systems control much of the music's character. The developments of recording technology, mentioned at the beginning of this introduction, also give us the backing track, offering vast sonic expansion but at the cost of significant musical control.

Moving forward a mere few decades takes us to the present day in the 21st century. Computer systems such as digital audio workstations (DAWs) and plugin hosts tie together a vast array of instruments and musical accompaniment options. A musical performer can now be a producer, conductor, and curator of high-level musical events, effortlessly shifting around arrangements and sections. However, all these options come at the cost of demonstrating direct control. This trend can be summarised as a general rule of music technology in performance events:

> The more sounds that an individual musician's setup can make, the less direct, real-time control the musician is likely to have over those sounds.

Our awareness of this relationship can help us to use music technologies and techniques to establish more exchange of energy in performance events.

I.2 TERMINOLOGY: EVENTS, EFFECTS, DEVICES, AND 'HOSTS/DEVICES'

The term 'events' is used here to distinguish between a musical performance in a recording studio (where the performances are often live at the time) and a musical performance at an event, such as a gig in a venue or a club. In a recording studio, the main aim is to play for a recording where there is not likely to be an audience. At an event, the main objective is to play for an audience. Accordingly, musical performance in these two different scenarios is quite different. In the recording studio, the performance focus is mainly on consistency, nuance, and as perfect a performance as possible. In an event, the focus is likely to be on demonstrating skill, flair, and exchange of energy.

The term 'effects' describes all sound manipulation tools, hardware or software. However, in Chapters 2 and 3, a distinction is made between effects and processors to establish workflow conventions for live sound and the recording studio (explained in Sections 2.2.3 and 3). For the rest of the book, the term 'effects' is used for both effects and processors.

The term 'device' is used as more of an umbrella term to describe any piece of software or hardware – either an effect, instrument, or utility. For example, a reverb unit is both an effect and a device, but a Musical Instrument Digital Interface (MIDI) keyboard controller piano is only a hardware device.

In the specific context of synchronisation between different equipment, the term 'host' is used for the system element that does the controlling, and the term 'device' is the system element that is controlled.

I.3 GENRES AND MUSICAL WORLDS

'Electronic music' is an umbrella term describing music created using instruments that work purely electronically (e.g., a synthesiser rather than an electric guitar). As is typical in Western music, electronic music follows the social divide between popular music and art music. In the case of electronic music, there is 'popular electronic music' and 'electronic art music'. Popular electronic music is likely to have the broad characteristics of a strong repetitive beat and standardised structure – for example, genres such as house and drum and bass.

Electronic art music tends to be generalised as electroacoustic music. It is likely to be more textural and less rigidly structured than popular electronic music. There is, of course, much overlap between the two worlds of popular electronic music and art music – for example, the seminal 1983 album *Apollo Atmospheres & Soundtracks* by Brian Eno. However, due to how music is marketed, distributed, and consumed, music that overlaps these two worlds tends to be the exception to the rule. While this distinction between popular music and art music is reductive, it is relevant for the different ways that the two worlds have made and continue to make use of technology, particularly regarding rhythm.

Popular electronic music grew from an approach that embraced the fixed tempo and repetitive nature of the sampler and drum machine. Electroacoustic music, on the other hand, has a far longer history and is generally based on the idea of sound art rather than being tied to time signatures and the 12-note systems of Western harmony. Due to this long academic history of sonic manipulation, much literature already discusses electroacoustic music. There is, however, relatively little literature on popular electronic music performance. Partly because this area is so overdue for exploration, this book focuses on live performance events of popular electronic music styles. However, this focus is not exclusively on popular music, as many techniques discussed are valid for both musical worlds.

I.4 TECHNOLOGY IN THE STUDIO AND ON STAGE

Technology for music production has developed to the point where once incredible and seemingly impossible processes are now commonplace. A multitrack recording can be captured, edited, and comped (the process of compiling edits of different takes) at incredibly high definition. It can be processed and sent through effects, using creative methods previously available in a limited fashion to only an elite group of well-funded sound engineers. Music can be crafted using electronic methods, building enormous and otherworldly sounds as editing and processing techniques are pushed in more experimental directions. All of this can be achieved using only a computer and an audio interface.

However, music production techniques are often significantly distanced from live musical performance. Once tracked, programmed (in MIDI for virtual instruments) and assembled (using samples/loops), it is standard practice for modern music to be created with clinical precision. This results in a sound that is vast and perfected in a way that is almost impossible to recreate in a live performance environment.

In contrast, live popular music performance events have maintained solidly reliable methods, often relying on nothing more innovative than a backing track. This has remained the case since the 1960s due to the overriding need for complete reliability on stage. The regularly demonstrated likelihood of a computer crash mid-set has been sufficient to deter all but the most resilient musicians.

However, computers and the software running on them have developed to the point that the incredible array of techniques once reserved for the studio and electronic music production is more and more accessible to the musician for reliable use in performance. The modern musician is now offered an array of sonic manipulation and augmentation that is daunting to even the most resistant to choice paralysis. However, all these techniques bring barriers to the exchange of energy in performance events.

I.5 CHAPTER OVERVIEW

This book presents the rudiments, best practices, and techniques to overcome the barriers to effective exchange of energy. Underpinning rudiments and best practices are divided into the following chapters: audio and live sound in Chapter 1; sound manipulation techniques of the recording studio in Chapter 2; electronic sound creation techniques in Chapter 3; and the tools of music technology in Chapter 4. Leading-edge concepts and techniques for augmenting performance events are explored in Chapter 5.

I.6 COST AND ACCESSIBILITY

This book leads with the broad underpinning practical concepts that apply to the technology most relevant to music performance in live events. Many specific examples are given throughout, using industry-standard technology, but these are by no means exclusive options. For example, the DAWs Pro Tools and Ableton Live are regularly referred to for offline music mixing and live performance tasks, respectively. Max for Live is also the programming environment used here. The cost of these tools may be prohibitively high, however. If so, it is worth researching the tools that best meet a performer's needs based on how accessible they are. This book does not offer a thorough list and review of alternatives; however, the relatively low-cost DAW Reaper and the no-cost programming environment Pure Data are worth exploring as suitable starting point alternatives. Where plugins are discussed, free options are mentioned wherever possible.

I.7 **NOTE ABOUT CATEGORISATION**

The concepts and techniques discussed in this book are separated into different categories (chapters and sections) for more clarity. Massive overlap exists between these concepts and techniques within practical applications of musical performance, music technology, and music production (as with most creative pursuits). Combining and juxtaposing other areas, disciplines, genres, and approaches often indicate innovative and interesting work. Hopefully, the categorisation used in this book gives a clearer understanding, which encourages rather than restricts the crossing of conceptual and technical boundaries and accordingly aids innovation.

ENDNOTE

1. Marr, J. (2016). *The Adam Buxton Podcast EP.51* [Interview]. https://www.adam-buxton.co.uk/podcasts/77

Audio and live sound

To make innovative and effective use of the live performance event environment, and to break traditional rules creatively and effectively without reinventing the wheel, it is necessary to know the fundamentals of sound, audio, and live sound. This chapter explores these fundamentals and establishes and explains the foundational best practices.

1.1 SOUND AND AUDIO

When making music, sound is our raw material. The more we know about the physics of sound, the more control we have over our sonic resources. This section explores the essentials of sound and audio.

To clarify the distinction between the two: sound is vibrations in the air that we can hear. Audio, by contrast, is recorded, transmitted, or reproduced sound – for example, a WAV file, the signal down a microphone cable, or sample playback.

1.1.1 Sound

1.1.1.1 Sound waves

At its fundamental level, sound is a mechanical longitudinal wave, which tells us two crucial things:

- Sound needs material to travel along – unlike light, which is an energy wave that can travel through the vacuum of space. The speed of sound in air is 343 metres per second at 20°C (while there are slight variations for humidity and air pressure, the significant factor is temperature), so it takes approximately 3 milliseconds (ms) (0.003 seconds) for sound to travel one metre. Accordingly, if we add 9 ms of pre-delay to a reverb effect, then our ears hear the start of the reverb as three metres 'further back', which makes the direct sound source appear closer to the listener.
- Sound is a pressure wave that moves lengthwise (longitudinally). If we consider the movement of a speaker cone as it moves back and forth, this compresses and refracts (expands) the air around it and transmits the sound to our ears. Figure 1.1 shows a two-dimensional representation of how sound waves travel away from the speaker cone for a tone with a single

DOI: 10.4324/9781003370406-2   **9**

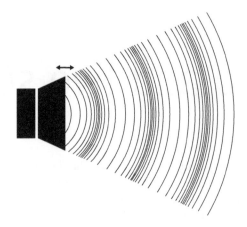

FIGURE 1.1
Two-dimensional representation of a loudspeaker and sound waves

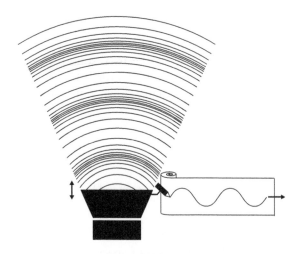

FIGURE 1.2
Loudspeaker pointing upwards and sine wave drawn as a scroll of paper unrolls to the right

frequency. The curved lines radiating out from the loudspeaker represent the compressed and refracted (expanded) air particles.

Even this simplification is a cumbersome way to visualise sound waves. Instead, we draw the displacement of air particles against time to give us the familiar image of a waveform. We can visualise this by putting the same speaker on its back and imagining an exaggerated vertical movement of the speaker cone, drawing on a moving piece of paper behind it. A pen in the figure then draws the familiar shape of a sine wave, shown in Figure 1.2.

1.1.1.2 Frequency

Whether the fundamental frequency of a pitched note (i.e. the pitch) or the frequency content of a complex sound (i.e., sound with several different frequencies), frequency is one of the most important characteristics of sound when making music.

The frequency of a pitched note is the number of times that note repeats in a second and is measured in hertz (Hz), or cycles per second. Every musical note has a fundamental frequency, otherwise known as pitch. It is worth noting that every sound other than a pure sine wave contains different, less prominent frequencies – called overtones – which make up its timbre.

The general hearing range of the human ear is 20 Hz to 20 kilohertz (kHz), but this is not a flat frequency response; different frequencies are heard at different volumes. The human ear is far more sensitive to the high frequencies shared by spoken consonants (5–7 kHz) than low frequencies (less than 100 Hz). See Figure 1.5, in Section 1.1.4, for a diagram showing the A-weighting curve – the most commonly used international standard that roughly approximates the human ear's frequency response.

Also, the human ear does not perceive frequency changes in linear, uniform steps, so this kind of relationship is referred to as a non-linear relationship. There are many non-linear relationships in sound and audio (e.g., volume, reverb, dynamic compression responses).

Linear relationships are widespread. For example, the amount (or volume) of something like water and its weight have a linear relationship. If you double the amount of water, it weighs twice as much; if you increase the amount of water by four times as much, it weighs four times more, and so on.

The frequencies of pitched notes, however, follow a non-linear scale. The relationship between musical notes and frequency does not change in amounts that are consistently the same (as water and weight do). The human ear hears octave intervals as having the same musical identity, so we can treat this as a large musical unit. For each additional octave increase, the frequency doubles rather than increases by the same unit amount as it would if the relationship were linear.

Figure 1.3 shows the frequencies corresponding to a piano keyboard starting on the note A0 (27.5 Hz) and going up to C8 (4186.01 Hz). The difference in frequency between two low-pitched notes an octave apart, A0 and A1, is 27.5 Hz. The difference in frequency between high-pitched notes an octave apart, A6 to A7, is a massive 1760 Hz. The human ear experiences the difference between A6 to A7 (1760 Hz) as the same musical difference of A0 to A1 (27.5 Hz). Also shown in Figure 1.3 are the approximate areas of low, mid and high frequencies using different shaded areas. Low frequencies are between 20 Hz and 200 Hz, mid frequencies are between 200 Hz and 3000 Hz, and high frequencies are between 3000 Hz and 20000 Hz.

When using equal temperament (the standard Western tuning system), the notes' frequencies between octaves are calculated as 12 even steps on a logarithmic scale. The logarithmic scale is one of the most common types of non-linear scales and is central to the concept of decibels (dB).

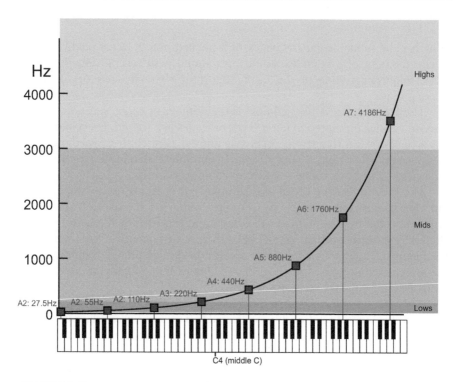

FIGURE 1.3
The frequencies of musical notes plotted against a piano keyboard spacing

1.1.2 Decibels

If a note's frequency is one of the most important sound characteristics for making music, then the decibel is one of the most important concepts in audio engineering. It is helpful to understand that the decibel is not a standard unit type, like a metre or a gram.

The decibel is a method of simplifying how we express massive ranges and describing the non-linear way that the human ear hears loudness (in a similar way to frequency, discussed in Section 1.1.2).

For example, the sound power range that the human ear can hear is measured in watts (W), and from the threshold of hearing to the threshold of pain is:

0.000,000,000,001 W to 10 W.

In dB sound pressure level (SPL), however, this becomes:

0 dB SPL to 130 dB SPL, which is significantly more convenient.

NB: dB SPL is dB sound *pressure* level, not sound power. Sound pressure and sound power are different acoustic units. The significant differences are that sound power is not dependent on the distance from the sound source, and its range is larger.

The dB, written on its own, describes an amount of change rather than any specific amount. This flexibility makes the dB particularly useful for faders on mixing desks, either physical or virtual, because the incoming signal that the fader affects is variable.

NB: For an analogue mixing desk, signal strength depends on both incoming signal gain and pre-fader effects. The signal strength for a digital audio workstation (DAW) depends on the audio file's amplitude and pre-fader effects.

So, in practical use on a mixing desk, a fader position of 0 dB means no gain change. A fader position of -20 dB will sound about one-quarter as loud, but in terms of power, is 100 times smaller. Figure 1.4 shows a screenshot of a fader in the DAW Pro Tools made by Avid. The 0 dB position is near the top, and the position for no sound is at the bottom and marked as infinity using the symbol ∞. All numbers under the 0 dB position are negative, but the negative markings are absent, allowing for a cleaner interface.

The letters 'dB' should be followed by other letters signifying what that amount is if they are to describe an amount of something specific – for example, dB SPL.

Three types of decibels are of particular interest for music:

- loudness (dB SPL and dB(A))
- voltage (dBu and dBV)
- digital (dBFS).

Generally, acoustic measurements in dB have a space (dB SPL), and electronic measurements in dB (dBu, dBV and dBFS) do not.

FIGURE 1.4
Screenshot of a volume fader in Pro Tools showing dB markings

1.1.2.1 Loudness (dB SPL and dB(A))

While 'loudness' is a subjective term, there are several reliable methods of measuring the objective loudness of a sound, the most common and straightforward
being dB SPL. dB SPL is the sound pressure in decibels relative to the quietest
sound we can hear (0 dB SPL). dB SPL is a good general measure of loudness,
but it does not take into account the frequency response of the human ear. To do
this, we need to use a weighted scale, which applies an equaliser (EQ) curve to
approximate the general frequency response of the human ear.

Figure 1.5 shows the A-weighting curve – the most commonly used international standard roughly approximating the human ear's frequency response. The
frequencies are shown in Hz on a logarithmic scale starting at 20 Hz and ending
at 20000 Hz, and the gain or reduction of sound level is shown in dB or decibels.

Like the pitches of equal temperament, the decibel uses a logarithmic scale. It
is worth noting that the quietest sound we can hear (the threshold of hearing) is
measured as 0 dB SPL. While it may seem odd that 0 dB SPL is an actual amount
of sound, decibels are the logarithm of a ratio against a reference. The reference,
in this case, is the quietest sound the human ear can hear. The ratio of the quietest
sound to itself, 1:1, is 1, and the logarithm (or *log*) of 1 is 0. Consequently, 0 dB
SPL is the quietest sound the human ear can hear.

It is also worth noting that while frequencies double for every octave, an
approximate doubling of loudness is achieved by increasing by 10 dB. Table 1.1
shows general dB SPL for sounds, threshold of hearing and pain.

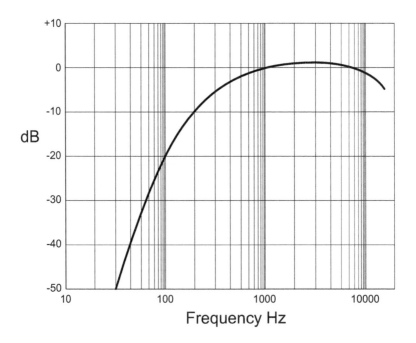

FIGURE 1.5
A-weighting curve

TABLE 1.1

dB SPL table

dB SPL	Description
140	Gunshot
120	Threshold of pain
115	Loud gig/club
105	Shouting
60	Conversation
30	Whisper
0	Threshold of hearing

1.1.2.2 Voltage (dBu and dBV)

In analogue audio, from mixing desks to synthesisers, the optimal signal level (i.e., the level that the equipment works best at) is called line level. There are two specifications of line level – one for consumer equipment and one for professional equipment:

- -10 dBV for consumer equipment
- +4 dBu for professional equipment.

dBV is based on the root mean square (RMS) (a measurement of average power) voltage relative to 1 volt. dBu is based on the RMS voltage relative to 0.775 volts over a 600 Ohm load.

What this means is that dBV and dBu are quite different. One specifies a type of energy with no load (dBV), and the other specifies a different reference energy with a particular load (dBu). The practical difference for equipment is that +4 dBu professional equipment runs at a higher voltage than -10 dBV consumer equipment. An essential measure of signal quality is the signal-to-noise ratio, which is the difference between the signal level and the noise. Because the signal level is higher at +4 dBu, the signal-to-noise ratio is likely to be better than -10 dBV for similar situations, and the relative noise level for +4 dBu equipment is likely to be less.

The lower signal level of -10 dBV means that components are likely to be less expensive, so -10 dBV equipment is likely to be less expensive. This classification of professional and consumer types of equipment extends beyond line-level considerations (allowing for the liberal use of the term 'professional' in sales hype), but this example is valuable for specific comparison.

As may be expected, a +4 dBu signal has a significantly higher level (is stronger or hotter) than -10 dBV. Some equipment may have switches or different inputs to allow operation at either level.

Another notable standard of monitoring level strength that has become ingrained in sound engineering is the VU meter. The VU meter remains popular due to the unique way the needle movement averages out the signal.

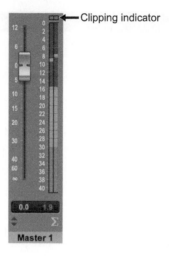

Clipping indicator

FIGURE 1.6
Pro Tools volume fader showing dB markings

1.1.2.3 Digital (dBFS)

dBFS stands for 'decibels full scale', meaning the full range of digital audio resolution. dBFS is the standard monitoring unit of most DAWs (not to be confused with fader position, which, like an analogue fader, indicates the amount of gain change). Figure 1.6 shows a view of a Pro Tools master (summing) fader expanded to show the level meter.

What this means in practice is that the absolute maximum level (at which clipping occurs) is 0 dBfs, and everything below is negative. The general rule for mixing in a DAW (especially with potentially high virtual instrument levels) is to keep each track monitoring at below -15dBfs, because each track that is mixed together will sum (add together and get louder). The rectangles at the top of the level meter indicate track clipping, which should be avoided – especially when recording – as clipping is a destructive and unpleasant distortion that is mainly impossible to remove. While in some systems, this distortion may not be audible (e.g., due to system headroom), it is still bad practice to run digital systems 'into the red'.

It is worth noting that digital clipping is very different to analogue clipping. Digital clipping is particularly harsh and, in nearly all circumstances, undesirable. (The clipping indicator in the Pro Tools monitor is pointed out in Figure 1.6.) In comparison to digital clipping, analogue clipping is likely to impart a warm, saturated sound which often suits musical styles such as classic rock.

1.1.3 Audio

As mentioned in Section 1.1, audio is recorded, transmitted, or reproduced sound.

Analogue audio: Analogue audio is sound turned into an electrical signal, allowing manipulation, transmission, amplification, and storage on magnetic tape or grooves in vinyl.

Analogue processes are vital at the beginning and end of the signal chain due to the analogue nature of sound, microphones, and amplification. Analogue signals have two key characteristics:

- They are continuous – that is, they are constantly changing.
- They have an infinite resolution between the minimum and maximum signal levels.

Digital audio: Digital audio is sound represented as a series of numbers (samples). An essential characteristic of digital audio is that it is a series of discrete steps. Digital audio is a sequence of finite values converted from an analogue signal. Once in the digital domain (i.e., a series of numbers rather than continuous values as in the analogue domain), the storage, manipulation, and transmission possibilities are typically far more varied and cost effective than their analogue equivalents.

The analogue to digital conversion process has two main steps: sample hold and quantising.

Sample hold: Because an analogue signal level constantly changes, the signal level must be held in place at one point in time to allow for effective measurement. This holding in place is done by a sample hold circuit, which holds the analogue signal level in place for the sample period, a duration of time set by the sample rate (discussed in Section 1.1.3.1).

Quantising: While the signal level is held in place, the quantiser (typically spelt 'quantizer', but the UK spelling is used here for consistency) measures the signal level and chooses the closest digital value.

The resulting output stream of samples is called pulse code modulation (PCM).

Digital to analogue conversion is necessary to turn the digital signal back into an analogue signal. This analogue signal can then be amplified and turned into sound energy by loudspeakers or headphone drivers for human ears to hear.

The process of digital to analogue conversion generally involves giving each digital value a correspondingly large analogue voltage and then smoothing the difference between successive sample voltages using a low pass filter (i.e., a filter that removes frequencies above a certain point) (filters are discussed in Section 2.11).

The analogue to digital conversion process indicates the two main factors that affect the quality of uncompressed digital audio (as well as the size of a digital audio file and speed of transfer): the sample rate and the bit depth.

1.1.3.1 Sample rate

The sample rate is the rate at which the analogue signal is sampled and is measured in Hz. The sample rate sets the period of the sample and hold circuit mentioned in the previous section. For example, a sample rate of 48000 Hz or 48 kHz has a sample period of 1/48000ths of a second (a number that rounds down to 0.02 ms).

The practical implication of the sample rate is that it determines the upper limit of the high frequencies that a digital system can capture, manipulate, and playback.

The sample rate must be more than twice the highest signal frequency to capture a signal accurately. This highest signal frequency, or 1/2 of the sample rate, is called the Nyquist frequency, named after electronic engineer Harry Nyquist. The sample rate for CD-quality audio is 44.1 kHz, so it has a Nyquist frequency of 22.05 kHz – a little over the general upper limit of the human ear, as mentioned in Section 1.1.1.2. Unfortunately, any frequencies that get through above the Nyquist frequency create errors in the captured audio, called aliasing. To stop aliasing, a low pass filter – which in this context is called an anti-aliasing filter – is used to remove frequencies above the Nyquist frequency.

Filters are discussed further in the following three chapters in the context of audio manipulation tools. An essential characteristic of filters in the context of digital audio system design is that they cannot immediately remove all frequencies beyond the cutoff without distorting the audio. The 'cutoff' is the point at which a filter starts to reduce frequencies by -3dB. The amount of frequency reduction after the cutoff is called the slope and is measured in a dB reduction per octave (i.e., 24dB/Oct).

With the previous sample rate of 44.1 kHz, the anti-aliasing filter is likely to reduce frequencies significantly lower than the Nyquist frequency of 22.05 kHz, so the anti-aliasing filter can be expected to remove some of the potentially audible, very high frequencies. Most professional recordings will be recorded with a sample rate of more than CD quality – generally either 48 kHz, 96 kHz or even 192 kHz. Sampling at high rates has the significant benefit of allowing an anti-aliasing filter with a much gentler slope so that higher frequencies can be recorded with less risk of distortion. Increasing the sample rate has the effect of increasing file size and processor load. However, it also *decreases* the time the computer takes to process and playback the audio. This time is called latency.

The considerations of latency and system stability are essential for live system reliability. While a stable system is an absolute priority, a latency of over 10 ms delay in monitoring is likely to be a noticeable annoyance to musicians. Over 20 ms will be almost impossible to work with; it is possible, but the monitoring will need to be ignored, making it pointless in the first place. The overall audio latency calculation is based on the number of samples processed at a time (called the buffer size) and the sample rate. Both buffer size and sample rate are set by the software processing the audio (i.e., a DAW). An initial fix for problems in sound playback, such as popping and glitching, is raising the buffer size in audio settings – for example, from 128 samples to 1024; but this will increase the latency. It is worth noting that round trip latency (input plus output latency), for typical DAWs used in the recording studio, is typically somewhere between 6 and 20 ms. For DAWs used for live performance events, latency will likely need to be as low as possible – ideally below 10 ms. For modern digital live sound systems, latency typically depends on plugin use, but can be as low as 1 ms.

1.1.3.2 Bit depth

Bit depth (or word length) is the size of the sample captured in 0s and 1s, which determines the number of values available for quantisation and, accordingly, the system's dynamic range. The dynamic range is the difference between the quietest and loudest signals the system can create and can be considered the system sensitivity. A 16-bit sample will be 16 digits long – for example, 0110 0100 0111 1100. A 16-bit 44.1 kHz audio file comprises 44,100 × 16-bit samples for every second.

The rule of thumb for bit depth and dynamic range is that each bit gives approximately 6 dB of dynamic range. A 16-bit audio system has appropriately 96dB of dynamic range. A 24-bit audio system has approximately 144dB of dynamic range. In the recording studio, it is best practice to record at 24 bits, whether exported audio will be reduced to 16 bits or not. The main reason for recording at 24 bits is that dynamic range compression, which is nearly ubiquitous in modern music production, raises lower-level signals. These signals will likely be cleaner and more detailed if recorded at 24 bits.

For a live system, however, it may be beneficial for system reliability to reduce the processor load and work at 16 bits for individual part capture.

1.1.3.3 Digital audio file types

There are three categories of digital audio files, each with different properties that lend themselves to particular applications. These categories are based on how (if at all) the data has been compressed – that is, made smaller.

First, it is worth understanding that there are several contexts for the term 'compression' in audio. In acoustics, sound is described as a compression wave, as discussed in Section 1.1.1.1. In the context of audio signal processing, 'compression' typically refers to dynamic range compression, where the peaks of a signal are reduced. Dynamic range compression is introduced in Section 1.2.3.2.2 in the context of live sound and in Section 2.1.2 in the context of the recording studio.

In the context of digital audio file types, however, 'compression' is used to describe the reduction in the size of a digital audio file, typically for ease of storage.

1.1.3.3.1 Uncompressed digital audio

Uncompressed digital audio is the output (typically PCM) from the analogue to digital conversion process. The audio on audio CDs is 16-bit 44.1kHz PCM.

For computer files, PCM is packaged into different file formats containing the information the computer needs to use the audio. The most common uncompressed digital audio file type is WAV, the standard default file type for PCs, which is typically compatible with most other operating systems. Apple's uncompressed digital audio file is AIFF. AIFF is identical in sound to WAV but is less compatible with operating systems other than MAC OS.

Most DAWs – such as Pro Tools, Cubase, Ableton and Logic – can record to either WAV or AIFF depending on operating systems, but in general, WAV is the more widely compatible of the two.

1.1.3.3.2 Lossy compression

Lossy compression reduces the file size of an audio file using psychoacoustic principles such as masking and critical bands to remove sound information less important to the human ear. The most common lossy audio file type is the MP3, produced by Fraunhofer Labs, which also created the AAC lossy format several years after. Both of these file types are proprietary files, meaning that there is a cost to use them. The most common non-proprietary lossy audio format is OGG, which works similarly to MP3 and AAC. At the time of publication, the streaming service Spotify uses OGG and AAC for their different players.

A one-minute CD-quality WAV file is just over 10 megabytes (MBy). If encoded into a 160 kilobits per second MP3, that file is reduced to just over 1 Mby. As the name suggests, data is lost in the conversion; so, while a new WAV file could be made with the MP3, it would have no more useful audio content than the MP3.

1.1.3.3.3 Lossless compression

Lossless compression reduces the size of an audio file using statistical algorithms so that the original can be reconstructed, giving no loss in audio quality. The most common types of lossless audio files are FLAC and ALAC. Like the more common general-purpose zip file/folder, it takes a significant amount of computer processing power to unpack or expand lossless files.

MP3 files, on the other hand, can just be played as they are, so early portable audio players did not support lossless files. A prevalent lossless format was FLAC, which had the benefit of not having any copyright protection (referred to as digital rights management). Because of this, however, there was little incentive for mainstream media player software such as iTunes to support FLAC.

It is worth noting that while many sample libraries are uncompressed, both Kontakt by Native Instruments and BFD3 by FXpansion use their lossless formats to save space when storing samples.

A lossless file will compress depending on the audio; however, if a one-minute 10 Mby CD-quality WAV file is encoded into ALAC, the new file will be about 5 Mby.

1.1.3.3.4 File type applications

Due to the high quality and speed with which they can be stored and manipulated, uncompressed files are the standard format for recording, sampling and live manipulation of audio – particularly for projects involving live sampling, resampling, looping and editing. Although the main drawback of uncompressed files is their relatively large size, technological advances in the size and speed of computer memory mean that this is less of an issue.

Lossy formats have been exceptionally popular for everyday music listening because there is little discernible sound difference to anyone other than an audio engineer or audiophile outside of a critical listening environment (e.g., using basic headphones). Based on the popularity of lossy streaming services such as Spotify, it is reasonable to conclude that most consumers do not seem concerned with the difference in sound quality.

That said, it is bad practice to use lossy files in producing audio for either performance, recordings or broadcast, because any initially minimal lossy artefacts will be accentuated by dynamic processing such as dynamic range compression, limiting and EQ.

TABLE 1.2

The general uses and relative pros and cons of different file types

Type	Example formats	Size for 1min 2track	Main uses
Uncompressed	WAV, AIFF	10MBy	Recording and production
Lossy	MP3, AAC, OGG	1MBy	General consumer listening, especially streaming
Lossless	FLAC, ALAC	5MBy (approx.)	High-resolution listening and sample banks

An overall comparison of the general uses and relative pros and cons of different audio file types is shown in Table 1.2.

1.2 LIVE SOUND

Live sound (also known as 'sound reinforcement') is the area of sound engineering that aims to project the sound of a performance to an audience and, if possible, to enhance that sound.

Apart from the quality of performance, the effective setup and use of live sound equipment make the difference between great and terrible sound. Modern electronic performers are likely to operate their equipment setup on stage. These setups need to be integrated into a live sound system called a public address system (PA) to be heard clearly.

It is worth mentioning that the pre-production of live performance is an often undervalued but vital phase for delivering a performance that sounds great. In live sound, as in studio recording, it is often all too easy to get lost in the technicalities of production. However, the most influential aspect of sound quality is performance. A good performance combined with effective live sound creates an environment for optimal exchange of energy between performer and audience.

1.2.1 The signal chain

A live sound system consists of a series of devices that take audio as an input, process it, and output it onto the next device. The following signal chain describes a standard series of analogue devices for achieving optimum live sound. While many modern setups use digital equipment between the stage box and amplifiers (i.e., digital mixing desk and effects/processors), most workflow is rooted in conventions set by analogue equipment, so an analogue setup is described here.

Each device is essential in ensuring sound clarity and system reliability, and can be separated into three groups based on use: sends, mixing, and returns. Figure 1.7 shows an analogue live sound signal chain split into these three groups.

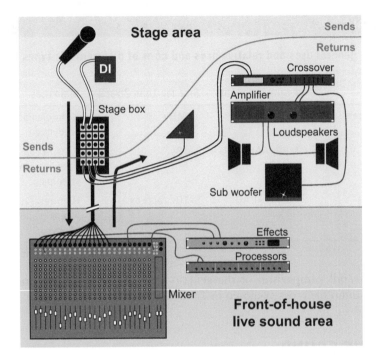

FIGURE 1.7
Analogue live sound signal chain

This signal chain exists within a standard performance event environment with two distinct areas: a stage and a front-of-house live sound area. The stage is where the music is performed and where the audio signals are captured. The front-of-house live sound area is where the audio signals are manipulated and mixed, typically in the middle of the audience.

This section introduces the fundamental elements that enable successful live sound: the different signal types; the cables and connectors used to link devices together; and the equipment used in the sends, mixing, and returns areas.

1.2.1.1 Signal types

In the context of a live sound system, what is meant by 'signal' is a representation of sound – either a stream of analogue electronic voltages or a series of digital samples. Acoustic sound energy is converted into electrical energy by a microphone (microphones are discussed in Section 1.2.2). This section discusses the different types of signals used by a live sound system.

1.2.1.1.1 Microphone level
Due to the relatively small amount of energy that acoustic sound has, the electrical output of a microphone is tiny – typically in the region of 0.005 to 0.05V. Because of this, mixing desks that take both mic and line inputs will have a mic/

line switch that changes the gain so that if the mic switch is engaged, there will be up to 60 dB more gain for that preamp.

1.2.1.1.2 Line level

Line level is used for electrically powered equipment, such as keyboards and samplers, where the energy available is due to design choices. As discussed in section 1.1.2.2, there are two types of line levels: -10 dBV for consumer (less signal strength) equipment and +4 dBu for professional (more signal strength) equipment.

1.2.1.1.3 Phono

Turntable output is different again from either type of line level. Standard turntable cartridge output is typically between 0.003 and 0.006V. As well as being lower than the mic or line levels, the frequency response is also different. To improve audio quality, vinyl records have the high frequencies boosted and the low frequencies cut (called RIAA equalisation). So on playback, the reverse EQ needs to be applied to the output; otherwise, the output will be tinny and lack depth.

1.2.1.1.4 High-Z

High-Z inputs are for sources with high impedance, typically guitar pickups. The output of pickups is between 0.1 - 1V, and while this is comparable to -10dBV line level, only a minimal amount of this voltage transfers into an input set to low impedance (i.e., mic or line). Hence, the signal may be faint, noisy, and lacking in high frequencies.

1.2.1.1.4 Digital audio

Digital transmission systems can provide a less noisy and more reliable format than analogue; however, the price of digital setups may be prohibitive for smaller setups.

The basics of digital audio transmission standards are as follows.

Professional two-channel digital audio transmission is set by the AES3 standard, typically connected by XLR or BNC connector. The consumer equivalent is SPDIF, connected by an RCA-phono connector. The main practical difference between SPDIF and AES3 is a shorter cable run allowed for SPDIF.

Multitrack professional digital audio transmission is often covered by the MADI/AES10 format allowing 64 channels. More modern network protocols, such as Dante or AVB, use an ethernet cable and can give over 1,000 channels of audio.

1.2.1.1.5 Digital control data/MIDI

The well-established standard for musical data is Musical Instrument Digital Interface (MIDI), which with most modern equipment, is connected using Universal Serial Bus (USB). The original MIDI 5-pin DIN connection is discussed in Section 1.2.1.2.2.

An alternate standard is Open Sound Control (OSC), which allows control over wireless networks.

1.2.1.2 Cables and connectors

While cables tend to be referred to by their use or connectors, a more detailed appreciation of critical factors is likely to be helpful. For example, a professional microphone cable is balanced. It typically consists of a shielded two-core cable terminated with XLR connectors: a female XLR connector on the output end and a male XLR connector on the input end. Balanced and unbalanced cables are explained below.

1.2.1.2.1 Balanced and unbalanced cables

Any length of metal will act as an antenna, picking up any electromagnetic energy crossing it. Cables are no exception; but unlike a radio antenna, all energy picked up along its length is unwanted – that is, it is noise.

Balanced cables have two separate cores and use differential signalling to cancel out most of the noise picked up on that cable. Differential signalling uses polarity inversion to cancel out any noise picked up by a cable length.

This process is best explained with the example of a vocal performance captured by a good-quality microphone. The microphone splits the signal in two and sends one unchanged signal down pin 2 (called the 'hot' signal) of the male XLR connection. It inverts the other signal's polarity and sends this down pin 1 (called the 'cold' signal) of the male XLR connection. Pin 3 is the shield or neutral element. Both signals travel down the two different cores of the cable, which pick up essentially the same noise.

At the next device (the preamp), the inverted signal is inverted again to return the vocal signal to its original polarity. Notably, the noise the cable has picked up now has its polarity inverted relative to the unchanged signal. These two signals are then added together, resulting in the noise disappearing and the vocal signal being twice as loud.

It is worth making a quick warning about terminology here: inverting the polarity of a signal is sometimes labelled 'phase invert' or 'phase reverse', but this is incorrect and somewhat misleading. A phase shift is a tiny *delay* proportional to a degree (one-360th) of the wavelength. A polarity inversion is when the positive and negative signal elements are swapped so that the top part of the graphical waveform is flipped with the bottom part. There is no delay involved.

So balanced cables for mono signals (only one signal) are cables with two cores and terminate with a connector with three connections (inverted signal, non-inverted signal and ground). Balanced cables most commonly terminate with an XLR connector and are often called a mic lead.

Unbalanced cables for mono signals terminate with a connector with two connections (signal and ground). Unbalanced cables most commonly terminate with a quarter-inch mono jack connector and are often called an instrument lead.

1.2.1.2.2 Connectors

The most common types of connectors (plugs) used in live sound and audio in general are described below and are shown approximately to scale in Figure 1.8.

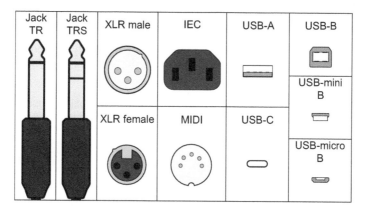

FIGURE 1.8
Common connectors

Three-pin XLR: The three-pin XLR connector (known informally as the XLR) is the standard connection for microphones and other balanced audio devices. There are two versions of the XLR connection: the male version with pins and the female version with sockets, allowing easy extension. In contrast to jack connectors, XLRs are more reliable, more secure (less likely to be unplugged by mistake), and easily chained together; but they are also bulkier and more expensive.

Quarter-inch jack: Unlike the XLR connector, all jack cables terminate in a male connector or plug. The female jack socket is used on the device or instrument. The jack connector is generally used for cost reasons, especially if the signal is high strength (as with the line outputs of piano keyboards) or if the cable is short (as is likely in home or project studios). The two types of jack most commonly used in live sound are the quarter-inch tip and ring (TR) jack and the quarter-inch tip, ring, and sleeve (TRS) jack. The quarter-inch TR jack – often called the mono jack or instrument lead – commonly connects guitars to amplifiers and powered audio devices (e.g., keyboards) to direct input (DI) boxes (to convert the unbalanced signal to a balanced one). The quarter-inch TRS jack is commonly used for unbalanced stereo connections – for example, headphones.

MIDI 5-pin DIN: Although USB connectors have mainly replaced the original MIDI 5-pin DIN connector, they are still occasionally used due to their rugged and reliable nature. Compared to the USB-B connection, the MIDI 5-pin DIN is more secure but is not powered and is only a one-way transmission. Because of this, legacy MIDI devices must use a dedicated power supply and separate MIDI cables for input and output.

USB (and Thunderbolt) connectors: USB is a standard for data transfer and uses several specific connector types. As the name suggests, USB is designed for various uses – most notably standard computer peripherals such as mice and printers. For audio, USB connectors are often used for consumer-level audio interfaces and MIDI devices. A USB cable typically connects a 'host' (e.g., a computer) to a 'device' (e.g., a MIDI keyboard).

TABLE 1.3

Dates and approximate transfer speeds of MIDI, USB1, 2, 3, and 4, and Thunderbolt 3 and 4

Standard	Year of release	Maximum approximate speed in Mbps
MIDI 1.0	1983	0.031
USB1	1996	12
USB2	2001	480
USB3.2	2017	20,000
USB4	2019	40,000
Thunderbolt 3/4	2016/2020	40,000

The year of creation and speeds of MIDI, the four versions of USB and Thunderbolt 3/4 are shown in Table 1.3. For devices with USB1 or USB2 standards, a variation on the USB A to B cable is likely to be used. The USB-A connector is rectangular, to be connected to the host (computer) end and one of the three forms of USB-B on the device end. The USB3 and USB4 standards are often used for devices requiring a faster connection and typically use USB C connectors on each end.

As well as USB, another notable connection for data transfer is Thunderbolt, an Apple Mac connection, which uses USB C connectors for versions 3 and 4.

As well as the approximate speeds listed, USB and Thunderbolt connections typically can provide low power levels and send input and output signals down the same cable.

IEC: Most standard audio equipment that does not have a transformer as a part of its power supply cable to mains will use the IEC 60320 C13 connector, often abbreviated to 'IEC'. The general terms 'computer lead' or 'kettle lead' are also used. There are many different types of other higher-rated power connections used in live sound and events, but these are outside the scope of this book.

1.2.2 Sends

Sends are the signals that come from the performance, typically on stage as shown in Figure 1.7. Sends can be categorised into three source types: microphones, line signals, and direct inputs. Microphones are discussed in Section 1.2.2.1. As both line signals and direct inputs are connected into DI boxes, they are discussed in Section 1.2.2.2.

1.2.2.1 Microphones for live sound

Microphones are essential for audio capture: they act as a transducer, changing the acoustic energy of sound into a voltage waveform that is analogous to the original sound signal. The word 'analogous' is noteworthy because it hints at an important fact: the signal coming out of the microphone is not the same as the acoustic signal. The design characteristics of a microphone will affect the sound it picks up and will impart a particular character (as our ears do). Some microphones have significant boosts and cuts in their frequency response, such as the live vocal microphone model Shure SM58, which has a frequency response intended to increase intelligibility. Other microphones have a flatter frequency response – for example, the studio reference microphone Earthworks M30, which has a frequency response that is as neutral as possible. Figure 1.9 compares a frequency response similar to the SM58 and the Earthworks M30. The microphone is the most critical of all of the signal chain devices. At best, a bad choice of microphone will cause the sound engineer to apply lots of corrective EQ. At worst, the microphone, PA, and monitor arrangement will be prone to howling feedback. While howling feedback is often referred to in live sound as just 'feedback', the distinction is made here because feedback has other uses and applications in broader audio applications – for example, the feedback of guitar amplifiers through electric guitars and the feedback in the design of filters in synthesisers (see Section 3.3).

A good microphone choice will give the most suitable sound for that instrument to fit into the overall mix, while rejecting spill and other potential causes of howling feedback. The main factors informing that choice are discussed in the following four sections.

Frequency response: Frequency response is the sonic character of the microphone expressed in the frequencies that the microphone accentuates or reduces. The frequency response of a microphone is typically shown as a graph in the frequency domain. Figure 1.9 shows a frequency response typical of a standard vocal microphone (e.g., a Shure SM58) and typical of a studio reference microphone (e.g., an Earthworks M30).

This graph shows that a frequency response similar to a Shure SM58 has a bass roll-off from 100 Hz and a strong presence and high-frequency boost

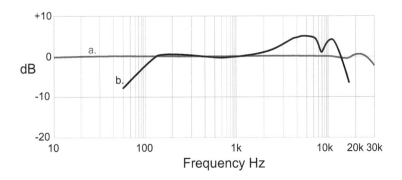

a. Frequency response similar to Earthworks M30

b. Frequency response similar to Shure SM58

FIGURE 1.9

Comparison between frequency responses of microphones similar to Shure SM58 and Earthworks M30

around the 5 and 10 kHz areas, which is consistent with the expectations of a vocal microphone.

Polar patterns: Polar patterns are circular diagrams describing how directional the microphone is, visualising the central axis of the microphone from above. Put another way, polar patterns show how sensitive a microphone is to sounds coming from different angles relative to the direction it is pointing, shown as 0° and up the page. It is worth bearing in mind that a microphone is likely to have different polar patterns for various frequencies.

Four standard polar patterns are shown in Figure 1.10. Cardioid microphones are the most common type for live sound because they are moderately directional, meaning they are better suited to picking up just one sound source. Cardioid microphones, such as the Shure SM58, will pick up sounds in front and a little to the sides, allowing for the sound source's movement without a significant signal loss. They also tend to give good rejection from behind the microphone, where a monitor speaker is likely to be. This good rejection is partly due to a particular design of the back part of the grill that takes in sound and uses it to cancel out howling feedback. Consequently, it is not recommended to cover up the back part of the grill – for example, by cupping the back of the grill when singing.

Hypercardioid (and similarly supercardioid) microphones offer greater degrees of directionality, giving better spill rejection but allowing less sound source movement. The bi-directional (also called figure of eight) pattern equally picks up sound in front and behind the microphone capsule. The omnidirectional pattern picks up from all directions, as the name suggests.

Maximum sound pressure level: The maximum sound pressure level, expressed in dB SPL, is the maximum sound pressure level that a microphone can pick up without distorting. The threshold of pain level for the human ear is roughly equivalent to the maximum sound pressure level of a microphone and is 120 dB SPL (although figures range from 115 to 140 dB SPL, reflecting the differences in human perception). The maximum span of the human voice measured about 2 centimetres from the mouth is 135 dB SPL, and the maximum SPL that a Shure SM58 can pick up without distorting is 160 dB SPL (at 1 kHz). Sound sources are likely to be significantly louder in live performance situations than in a recording studio, so it is essential to know that the microphone will not start to distort, especially as the performance is likely to be played louder than the soundcheck.

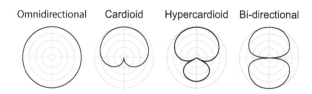

Omnidirectional   Cardioid   Hypercardioid   Bi-directional

FIGURE 1.10
Four standard polar patterns

1.2.2.1.1 General microphone placement

Getting the correct sound is much more a case of critical listening and experimentation than following any rote live sound dogma. That said, there are best practice techniques worth knowing to avoid reinventing the live sound wheel. These techniques are informed by an awareness of the construction and playing mechanics of the instruments being miced up. For any new or unknown instrument, it is recommended to study how the instrument works and how it is played, and apply the best practices discussed below. Specific instrument best practices are discussed in Section 1.2.2.1.3 for dynamic microphones and Section 1.2.2.1.5 for condenser microphones.

For every sound and space, there are one or more 'sweet spots' where the instrument's (or amplifier's) sound is best suited to the overall needs of the music. Sound engineers will know where these sweet spots may be or will check using their ears (if the instruments are not too loud). Once these ideal positions are established, the need for isolation from other sound sources can be considered.

In most live sound event cases, isolation can be achieved by keeping a microphone with good spill rejection close to the source and keeping other sound sources as far away as realistically possible. However, it is sometimes advantageous to make use of a barrier to the sound – for example, a clear shield around a drum kit.

Distance and direction to the sound source go hand in hand, and different microphones will react differently depending on the instrument, but the basic rules of thumb are as follows.

The most obvious rule is that the sound gets louder the closer the microphone is placed to the source, but the sound is also likely to get harsher and more direct. As well as this, most microphones tend to accentuate bass frequencies as the microphone moves closer to the sound source, which is called the proximity effect (this is partly why rock singers tend to 'kiss the mic').

As discussed in Section 1.2.2.1, the direction of a cardioid microphone relative to its central axis (the direction it is pointing) makes a significant difference to how much it picks up, but it also affects the signal's tone. A microphone that is 'on axis' is pointing directly at the source (90°) and will tend to sound bright. If a microphone is 'off axis', it is not pointing directly towards the sound source and its angle relative to the sound source will be greater than 0°. Typically, the more off-axis the microphone is, the fewer high frequencies it will pick up. It may also be worth experimenting with the microphone pointing away from the sound source, allowing it to be closer, but more off-axis will typically give a louder but warmer sound.

1.2.2.1.2 Dynamic microphones

The core components of a dynamic microphone are a magnet inside a moving coil connected to a diaphragm. Sound waves hit the diaphragm, which moves the magnet and generates a tiny electrical signal in the moving coil. Because of the mechanical nature of dynamic microphones, they tend to be rugged and reliable, although they are not as sensitive or detailed as condenser microphones. As well as this, dynamic microphones do not pick up much distant sound; in a live environment, this is especially advantageous, as this is another way of saying that they are good at rejecting spill or bleed (i.e., the sound from other instruments).

Because of this, in a live sound environment with a large PA, most microphones will likely be dynamic rather than cardioid, as they are less prone to howling feedback than condenser microphones and tend to have high maximum dB SPL levels.

Most microphones will be made to suit a particular type of sound source – for example, vocals, instruments, and bass instruments. As discussed in Section 1.2.2.1, vocal microphones such as the standard Shure SM58 have a frequency boost in the presence peak, which can be suitable for vocal clarity. Vocal microphones are also likely to have good rejection from behind the microphone, making them less prone to howling feedback from a monitor.

Instrument microphones are likely to have a flatter, less coloured frequency response than vocal microphones, allowing for a broader range of instrument types. Microphones for bass instruments, such as a kick drum or bass guitar amplifier, will have a lower bass cutoff and potentially a boost in the bass frequencies, such as the 100 Hz boost of an AKG D112.

1.2.2.1.3 Dynamic microphone placement

Due to the need to reduce spill and the risk of howling feedback, most dynamic microphones used live are 'close miced' – that is, they are rarely more than ten centimetres away from the sound source.

Dynamic microphones are typically placed as close to the source as possible as a general starting point without getting in the performer's way. However, as well as the general microphone best practice discussed in Section 1.2.2.1.1, there are the most common, specific instruments that have different best practices: kick drum, snare drum and toms, guitar amplifiers, and vocals.

Kick drum: Most drums consist of two drum skins, called heads: a batter head, which is hit; and a resonant head, which – as the name suggests – resonates. Both heads are held in place by a drum shell. The kick drum is also called the bass drum, but if abbreviated to 'bass', this can be confused with a bass guitar, so the term 'kick drum' is used here. Typically, the microphone is placed inside the shell through a hole in the resonant head. The closer the microphone is to the beater (the part of the pedal that hits the batter head), the more the sound will tend to have mid-frequency content. As the microphone moves away from the batter head, the sound will typically get warmer until the microphone leaves the shell and points at the resonant head. At this point, the more boomy, low frequencies of the resonant head will be picked up. As the microphone leaves the drum shell, there is a greater risk of picking up sound from other sources.

For more elaborate setups where enough channels are available, it may be helpful to use multiple microphones on the kick drum – that is, an inner microphone placed close to the batter head and an outer microphone just at the hole of the resonant head. These sounds can then be mixed to achieve the tone required.

Snare drum and toms: For snare drums and toms, there is rarely access inside the shell for microphones, so the microphone will tend to point down at the batter head. Depending on the drum, microphone and drummer playing style (it is best not to hit the microphone…), the microphone will point somewhere between the centre of the head and the rim.

The snares of a snare drum are metal ribbons touching the resonant head that cause its distinctive sharp attack and harsh tone. As with the kick drum, where enough channels are available, it may be possible to have a microphone on the top (the batter head) and a microphone on the bottom (the resonant head), allowing the different tones to be mixed. Because the top microphone is pointing down and the bottom microphone is pointing up, the polarity of the bottom microphone will be opposite to the top microphone, which will result in much of the sound cancelling out, and which is likely to sound weak and thin when combined. If the bottom microphone has its polarity inverted (most mixing desks have a switch with this function bearing the Ø symbol), this is likely to solve the issue.

Guitar amplifiers: Although the amplifier is technically the electronic device that increases the signal level, when used generally for live sound, the term also includes the cabinet containing the loudspeaker cones. In cabinets with more than one cone, these tend to be the same type; however, this is not necessarily the case, and it is worth finding out whether there is a particular cone that is best for close micing. The part of the cone that the microphone is pointing at and the microphone's angle will also make a tonal difference. Essentially, the more on-axis (facing straight at the amp) the microphone is, the brighter the tone will be; and the more off-axis the microphone is, the warmer or deeper that tone will be.

Vocals: Optimal microphone placement for vocals depends on how directional that microphone is, how pronounced its proximity effect is, and how that suits the singer's technique. It is worth discussing how much movement the singer wants to make and how much they want to use a microphone stand. The general issue with underconfident singers is that their singing volume is likely to be relatively low, and they may not be close enough to the microphone, so the channel gain needs to be much higher, which increases the risk of howling feedback.

1.2.2.1.4 Condenser microphones

Condenser microphones (also called capacitor microphones) use two thin plates of metal spaced closely together, one of which is connected to the microphone's diaphragm. Sound waves hit the diaphragm, which changes the distance separating the plates and creates an audio signal. Because the moving plate of the condenser microphone is so much lighter than the magnet of a dynamic microphone, they are significantly more sensitive to a wide (in particular, high) frequency range and transients (quick changes in energy). Condenser microphones, unlike dynamic microphones, need to be powered using phantom power – 48 volts of power typically supplied by the preamp.

Because of their sensitivity and relatively low maximum dB SPL levels, condenser microphones are not widely used for live sound as they are so prone to howling feedback. Two exceptions to this rule are a drum kit's overhead microphones and vocals. For drum kit overhead microphones, a more robust and less sensitive type of condenser will typically be beneficial – for example, an AKG C1000. For vocals, while condenser microphones are not as widely used as dynamic microphones, the sensitivity can make up for the comparative lack of feedback rejection. Vocal condenser microphones tend to be significantly more expensive – for example, the Shure KSM9 live vocal microphone is about five times the price of the dynamic Shure SM58.

1.2.2.1.5 Condenser microphone placement

Due to their sensitivity, the primary consideration for condenser microphone placement is how best to minimise the risk of howling feedback. This risk can be minimised by pointing microphones away from monitors and keeping microphones as close as possible to the sound source. As the distance from the sound source(s) increases, as with drum overhead microphones, it is helpful to consider how reflective or absorbent the space is. If the room is lively, it is worth placing some absorbent material behind the performance area if possible.

For tonal considerations, the general microphone best practice discussed in Section 1.2.2.1.1 applies here, as a cardioid pattern will likely give the best isolation. As condenser microphones provide more opportunities to explore distance micing techniques, it is advantageous to try various approaches and analyse them with a critical ear. It is also worth being aware that a space with no audience (e.g., in soundcheck) will sound very different from a room full of people. Typically, people are good absorbers of high and mid frequencies.

1.2.2.2 DI boxes

DI boxes are designed for direct input of both line and electric instrument sources. Their primary purpose is to change an unbalanced signal into a balanced signal to allow for longer cable runs with less noise.

Line signals are signals from electronic equipment such as synthesisers, on-stage mixing desks (for turntables or multiple electronic instrument setups), or laptops. For more detail about signal types, see Section 1.2.1.1.

Electric instrument sources tend to be either passive (unpowered) instruments with magnetic pickups, such as guitars or bass guitars; or low-powered instruments with piezoelectric pickups, such as electro-acoustic guitars or violins.

There are two main types of DI boxes: passive (unpowered) and active (powered by mains, battery or phantom power). The general rule of thumb is that it is best to connect an active source (e.g., a synthesiser) into a passive DI box and a passive source (e.g., a bass guitar with no battery) into an active DI.

1.2.2.3 Stage boxes

In a standard live sound setup, the microphones and DI outputs are the sends and are connected to the stage box, as shown in Figure 1.7. Also connected to the stage box are the desk outputs for the main speakers and monitor mixes, which are the returns.

A stage box typically comprises several rows of female XLR connectors for the sends and fewer rows of male XLR connectors (or jack sockets on cheaper models) for the returns. A standard analogue stage box is attached to a large, heavily protected multicore cable, or 'snake', that takes each signal to the mixing desk. The digital equivalent to a large, heavy multicore is typically an ethernet cable, allowing a much easier setup and less noise interference.

For large performances with big enough budgets, there is the option of spitting the sends to the front-of-house (FOH) mixing desk and a monitor mixing desk. This separation allows the FOH sound engineer to concentrate on the mix that the audience is hearing. Managing the performers' monitor mix then goes to a monitor sound engineer, who will typically be set up on one of the far sides of the stage.

1.2.3 Mixing

Mixing is all the sound manipulation processes of optimising and combining individual instruments/channels into a mix. There are typically two mix types in live performance events: the front-of-house mix and the monitor mixes. The front-of-house mix is the sound that the audience will hear and the monitor mixes are for the performers. Typically, monitor mixes are played through monitor speakers (floor wedges or side fills) or in-ear systems.

In creating a front-of-house mix, the sound engineer manipulates each channel (typically, each channel is a different microphone or DI box) so that the sound fits into the mix in a way that serves the musical aesthetic. Each channel should be heard clearly and play a part in the overall sound of a mix without drowning out or detracting from the other parts. A helpful model for encouraging good mix practice is picturing the mix as a sonic soup. Each instrument is like each ingredient, which, when cooked (i.e., mixed), effectively plays its part – some prominent and others combining to create a particular result.

The tools that a sound engineer has to manipulate each channel are a variety of processors and effects. There is a division between processors and effects as sound manipulation tools because, in standard practice, they are used in significantly different ways, which may be summarised as follows:

- Processors are hardware devices inserted into the channel. They are typically used to remove or replace audio elements – for example, an EQ.
- Effects are hardware devices that have their dedicated channel. They have signals sent to them, as these sounds are typically added to the original signal – for example, a reverb effect.

This division is explained further in the following section and should not be seen as a hard and fast rule – more good practice for regular use and a departure point for innovation. The fundamental principles and differences between processors and effects are explained here and elaborated on in the context of live sound in Sections 1.2.3.2.2 and 1.2.3.2.3. The use of processors and effects in the recording studio is discussed in Chapter 2.

Processors: Processors are devices that change a signal's dynamics or frequency content. As mentioned above, for regular use, processors are inserted into a channel – that is, they break the flow of the signal and, if turned off without bypassing, will silence the channel. There are two main types of processors: dynamic processors and EQs.

Dynamics are the varying volume levels of a signal. A very dynamic signal will have a significant volume difference between the loud and quiet parts, while a less dynamic signal will have a more consistent volume. In summary, dynamic processors change a signal's varying volume levels. Typical examples of dynamic processors are dynamic range compressors (simply called compressors in this context) and gates.

EQs are devices that change (or process) a signal's frequency content or tonal characteristics.

Effects: Effects add elements such as delay, reverb or distortion to the signal. Standard practice is to give each effect its dedicated channel (called the wet signal), and send a proportion of the original signal (called the dry signal) to the effect channel. This wet signal is then mixed into the main mix along with the dry signal.

However, using processors as inserts and effects as sends is not a hard and fast rule. It is more of a starting point for good practice and something to be aware of for effective creative use when it comes to more experimental approaches to these tools.

While mixing for a live performance is based on the same techniques as mixing in a recording studio, the approaches and skillsets needed are quite different in practice. The main reasons for these differences are that a live sound engineer does not have the editing environment or the additional time of the studio engineer. Consequently, processors and effects used in a live context may do the same job, but they need to be easier to set up and operate to achieve optimal results quickly and in real time. Consequently, processors and effects for live sound tend to lack the complexity of recording studio processors and effects.

1.2.3.1 Live mixing desk

As with effects, equipment for live events typically needs to be quicker to set up and operate than recording studio counterparts, and the live mixing desk is a prime example of this. A live mixing desk is typically optimised for rapid real-time audio manipulation and problem-solving. In contrast, a mixing desk in a recording studio is generally optimised for clinical, highly accurate offline audio manipulation and enhancement.

A traditional analogue mixing desk mainly consists of multiple identical channels: columns of knobs and buttons to control each incoming signal. Typically, there are between eight and 32 channels for signals from the stage (the sends) and effects. Also, there are between four and eight group channels and a master channel to the right of the desk. Digital mixing desks tend to have fewer dials and buttons per channel as they are multi-functional, and are likely to have screens (often touch screens) to manage routing, grouping processors, and effects that run internally.

As digital mixing desks tend to be modelled on analogue mixing desks, it is worth understanding the rudiments of an analogue desk and the routing of analogue outboard devices. Outboard devices are physically external to a mixing desk or DAW, typically rack mount processors or effects.

The main controls on a standard analogue mixing desk from the top indicate the signal chain through the desk. Figure 1.11 shows a simplified diagram of a

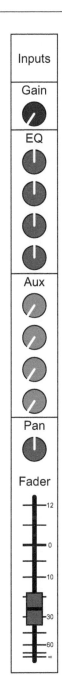

FIGURE 1.11
Simplified channel strip of a typical live sound mixing desk

typical channel strip on a live sound mixing desk. As with most equipment, there is significant variety in operation and features. The sections below describe the typical elements of a live mixing desk set out to best explain its rudiments.

Inputs and inserts: The input and insert sockets are usually at the top of a live mixing desk. The inputs are typically a female XLR microphone input and a quarter-inch jack line input. There will also be a quarter-inch jack insert socket on most mixing desks, which allows devices – typically processors (e.g., a gate) – to be inserted into the signal of the mixing desk.

Gain: The preamp gain (also called mic trim) is a single rotary knob that sets the level of the signal to be at the optimal strength for the mixing desk to give the maximum amount of volume control for the fader. It does a similar job as the preamp to gain control of an audio interface.

EQ: Equalisation consists of a series of filters that adjust the signal's frequency content. EQs and filters are fundamental elements of sound manipulation. Section 1.2.3.2.2 discusses filters in the context of live sound, Section 2.1.1 in the context of the recording studio, and Section 3.3 in the context of sound synthesis. On a live mixing desk, there are typically between three and seven rotary knobs for either shelving or sweepable filters.

Auxiliary sends: The auxiliary sends (aux) are controls that set how much of the signal on that channel is sent to what is commonly an effect or a monitor mix. There will typically be between three and seven aux controls with a single rotary knob, each controlling the level sent.

Pan: The pan pot (short for potentiometer) knob, or panning control, controls how much of the signal goes to the left or right channels of the main mix.

Fader: The channel fader on a live mixing desk is the primary control for that signal's level. The track design of the fader allows it to be operated (or 'ridden') with maximum ease and accessibility. As with the DAW fader shown in Figure 1.4, live faders typically go from minus infinity -∞ at the bottom to about +12dB at the very top. At the 0dB point, the fader makes no change to the signal level (so volume at this point is solely set by the gain control), and is about four-fifths up the fader track.

1.2.3.2 Live event sound manipulation tools

As stated in Section 1.2, live sound is the area of sound engineering concerned with projecting the sound of a performance to an audience and, if possible, enhancing that sound. In live sound engineering, these sonic improvements are typically corrective rather than creative because a sound engineer's traditional focus is to project the sounds of the musicians authentically. As this chapter deals with live sound rudiments, the focus is on the corrective use of sound manipulation devices. More creative approaches are explored in Chapters 2, 3 and 5. The difference between corrective and creative approaches to sound manipulation is explained in the following section.

1.2.3.2.1 Corrective and creative sound manipulation

A corrective use of sound manipulation devices – either processors, effects, or both – is an approach that uses these devices to correct sonic issues transparently – that is, the sound's character will not be significantly changed. For example, there will likely be lots of noise between the hits on a close-miced snare channel. This noise may form a bed of sound that reduces the mix's clarity. A gate inserted into that channel can remove the low-level noise (gates are explained more thoroughly in Section 1.2.3.2.2).

The gate's result is unlikely to be apparent to the audience, but the overall mix is likely to sound cleaner. The cumulative effect of many well-managed corrective processors and effects is to create a significantly more engaging live sound experience for the audience.

Creative use of sound manipulation devices is a more conspicuous use of either processors, effects, or both to change the character of the sound – often dramatically and conspicuously. For example, a noise gate inserted into the channel of a textural synth pad may be triggered by a different signal/instrument – a technique called sidechaining, which is discussed further in Sections 1.2.3.2.2, and 2.1.2.4. If this sidechain input has a particularly percussive character, the previously smooth sustained sounds of the synth pad will now be aggressively choppy sounding and have a significantly different feeling.

There is, of course, a middle ground between purely corrective and purely creative approaches to sound manipulation. Depending on the situation, it is worth considering *how* corrective and creative the use of each device should be. This balance of corrective and creative approaches is particularly relevant when considering recording studio and electronic music environments (discussed in Chapters 3 and 4, respectively).

1.2.3.2.2 Processors

As introduced in Section 1.2.3, processors fall into two categories: dynamic processors and EQs. Dynamic processors change the dynamics (the differences in volume), and EQs change the frequency content. Processors are perhaps best understood as solutions to common live sound problems in the form of equalisers, compressors, and gates.

EQ: If the sound of an instrument is too dull or too bright, an EQ can be used to change the frequency content of the sound. On all but the most basic analogue mixing desks, there will tend to be an EQ on each channel (as opposed to compressors and gates, which tend to be outboard on analogue desks). As shown in Figure 1.11, the EQ controls on a mixing desk will typically consist of the following:

- a high-frequency shelving filter
- two sweepable mid-frequency band pass/reject filters, and
- a low-frequency shelving filter.

High-frequency shelving filters use a single rotary knob to either boost or cut frequencies above a fixed frequency – typically around 12 kHz. Low-frequency shelving filters boost or cut frequencies below a fixed frequency – typically about 80 Hz. Shelving filters allow for quick and broad changes to the tonality of a sound – for example, reducing the low frequencies that may be causing rumble or boosting the high frequencies of a dull-sounding instrument.

Sweepable bandpass/reject filters use two rotary knobs: one controls the centre point of the filter in Hz, and the other controls how much boost or cut is applied to these frequencies in decibels. This frequency control allows the engineer to quickly identify a particular set of frequencies by boosting and sweeping the filter across the frequency range. As the filter is swept across its range with a boost setting, it produces a swooshing sound as it picks out the different frequencies on the channel, allowing identification of the frequencies that need boosting or cutting. A channel EQ typically has high, mid, and low-mid filters. These allow the typical boost at the presence peak, around 3–5 kHz, to add brilliance to the sound, and cut at 200–300 Hz – the area most likely to cause a muddiness to the sound.

A simple high pass filter button is the other widespread EQ control on a desk, which reduces frequencies below a certain point – generally around 100Hz. The button tends to be close to the gain control and is separate from the EQ, so the high pass filter will still be active if the EQ controls are bypassed.

Compressors: If a singer has a wide dynamic range, the typical problem for a live performance event will be that the loud parts are too loud and the quiet parts are too soft. While it is possible to ride the fader to balance these level differences, this is a very challenging task requiring uncanny prediction and consistent focus, making it unrealistic in a live event situation. (Riding the fader is discussed in the context of the recording studio in Section 2.1.5, where it has a similar but more influential role.) A more common, reliable, and consistent solution is to use a compressor.

A compressor is a device that reduces the signal's strength over a particular threshold by a ratio. For example, the threshold of a compressor is -30 dB, and the ratio is 4:1. As the signal goes over -30 dB, only one-quarter of the increase is output – for example, if the input signal is 4dB over the threshold, only a 1 dB signal increase will be output. The immediate result is to reduce signal strength, so a make-up gain control is added after the compression to boost the signal, making the quiet parts louder. The practical outcome of using compression will tend to make the overall signal sound bigger.

Gates: The problem of a channel with unwanted noise in periods of no sound was discussed in Section 1.2.3.2.1, with the example of a close microphone on a snare drum picking up lots of noise between hits. This noise typically comes from other sounds on stage (so it is called spill), and is likely to have a lower signal level than the snare hits. The gate (or noise gate) works by only letting sound through with a higher signal level than the noise – otherwise known as the threshold – in dB, at which the gate will open. Attack and release are the other main controls that affect the gate's operation. The attack controls how quickly the gate opens. A faster attack will sound sharper and allow for a harsher initial transient; and a slower attack will smooth out the attack of the note or hit. The release (which can also be combined with a hold setting – that is, how long the gate is held fully open) defines how quickly the gate closes. A faster release gives a more choppy and percussive sound, and a slower release lets through more of the tail end of the sound, giving a more sustained sound. It is worth noting that gate plugins in DAWs can be faster than outboard equipment, as plugins can use system delay to look ahead at the incoming audio. (Plugins are additional software programs that run within a host program, allowing expanded capabilities.)

1.2.3.2.3 Effects

As introduced in Section 1.2.3, effects are devices that add variations of the signal (the wet signal) onto the original (dry) signal. The most common examples of effects are reverb and delay, which are explained here with a solution-based approach as in Section 1.2.3.2.2.

Reverb: As discussed in Section 1.2.2.1.2, most microphones in a live environment are dynamic microphones positioned close to the sound source so that the signal is as clear and as strong as possible, and with as little spill as possible. Using close dynamic microphones also means that the sound picked up is likely to lack any acoustic characteristics of the room.

The human ear is accustomed to hearing sounds in the context of a space. Sounds bounce around an enclosed room, creating many tiny delays – or reverberations – that all combine to give the sound character and add important spatial information about where that sound is and what is near it. Without these reverberations, the direct sound would likely be isolated and dry.

A reverb effect takes the dry sound and attempts to replicate the sound of an enclosed space by adding these tiny delays. The primary reverb control reflects the main acoustic measurement of an enclosed space: the time it takes for the reverb, or slight delays, to die out. A short and tight-sounding reverb will be below two seconds, and a long and lush-sounding reverb will be above two and a half seconds, depending on the methods and characteristics of the reverb device. Other controls likely to change the character of the replicated space include room size and EQ to change the tonal characteristics of the wet signal.

Before digital systems, the two leading analogue solutions to dry signals were plate or spring reverb. These mechanical devices played sound into an object – typically a metal plate or spring – which would vibrate as the waves bounced around. The resulting combination of these waves was picked up somewhere else on the plate or spring, giving a reverberant signal. Because of the mechanical nature of these devices, they have very characteristic sounds. A plate reverb tends to be quite harsh and sharp, so it can be well suited to a direct snare signal. Spring reverb has a particularly boingy sound and is extremely popular with guitar amplifiers. Because of how integral these devices were to sound recordings and their enduring usefulness, plate and spring are typically settings on modern digital reverb units alongside different enclosed spaces, such as rooms, halls and churches.

Delay: Delay is another method of adding character to a signal. As the name suggests, delay takes the original signal and delays it for an amount of time, adding to the original signal. This time is typically measured in milliseconds or seconds. A variation to using seconds or milliseconds to control delay time is synchronisation or sync mode, where rhythmic values set the delay time relative to the tempo to add more percussive effects that are always in time. Typical settings include ⅛, ¼, or dotted ¼ notes, which would be 250, 500, and 750 ms at 120 beats per minute.

Feedback control is also often included in delay effects, which feeds the delayed signal back into the input of the delay effect, causing textural layering at low settings to a cascade of noise at high settings.

Compared to reverb, delay is a more discrete method of adding character to a dry signal, as there are far fewer delays than reflections. A combination of both reverb and delay is often highly effective.

1.2.3.3 Mixing live

As discussed in Section 1.2.3, mixing is the optimising and combining of individual instruments/channels for the front-of-house or audience mixes. An essential characteristic of a good mix is correct balance – that is, where the relative levels of the instruments and vocals are adjusted so that no element is too loud or too quiet. Standard practice is that the processors, effects, and initial balance will have been set up and tested in the soundcheck before the performance. The sound engineer is then left with the task of making alterations to the sound as the performance progresses and, if necessary, solving any problems as they arise.

As discussed in Section 1.2.3.2, the traditional and typical role of the live sound engineer is to make corrective rather than creative alterations to the sound, which is likely to be a continual refinement of the overall balance as the performance progresses.

1.2.4 Returns

As shown in Figure 1.7, the front-of-house mix is typically returned to the stage down the multicore or snake (along with any monitor mixes if there is no separate monitor mixing desk). These signals are at line strength and will typically be sent down XLR cables to be amplified and projected to the audience (or the performers in the case of a monitor mix). The following chapters describe the different devices that make this happen.

1.2.4.1 Power amplifiers

There are two types of amplifiers for audio signals: pre-amplifiers and power amplifiers. The pre-amplifier is the first part of the mixing desk. It is controlled by the gain control shown in Figure 1.11 and brings up the signal to line strength (discussed in Section 1.2.1.1.2), so that it can be better manipulated using effects and processors. Line strength is far too small a signal strength to drive a loudspeaker, so the signal needs to be amplified by a power amplifier.

Typically, PA power amplifiers are two-channel devices with a rotary knob to control the amount of amplification for each channel. Most power amplifiers can work in a two-channel mode for separate left and right channels or can be bridged to output a single channel (mono) at twice the power.

The most straightforward arrangement of a separate amplifier and loudspeaker is to connect the output of each amplifier channel directly into each loudspeaker (left and right), typically using speaker cables.

1.2.4.2 Loudspeakers

Loudspeakers, commonly abbreviated as 'speakers', are the final link in the signal chain. The core components of loudspeaker drivers are a coil of wire surrounding a moving magnet attached to a cone. As an electrical signal passes through the coil of wire, it moves the magnet and the cone. The cone then pushes the air in front of it air back and forth, creating acoustic sound energy. It is worth noting that the mechanism of a loudspeaker is essentially a dynamic microphone in reverse, which – neatly enough – is the first link in the signal chain.

The simple setup described previously in Section 1.2.4.1 is common in small setups, meaning that the loudspeakers must output the full range of frequencies.

Loudspeaker cones are typically mounted in large cabinets with other loudspeaker cones, each taking a different range of frequencies. Large cones are better suited to creating low and mid frequencies (due to the longer wavelengths). Small cones are better suited to creating high frequencies and are often coupled to a horn to increase efficiency and make the sound more directional.

1.2.4.3 Crossovers

Larger setups are likely to use different amplifiers and loudspeaker combinations for different frequency ranges. The device that splits these frequencies is called a crossover. The most straightforward split (illustrated in Figure 1.7) would be between high and low frequencies. In this case, the front-of-house mix is sent to a bi-amp crossover – essentially two filters that split each channel into high-frequency and low-frequency audio. The primary control for the crossover is the frequency at which the split takes place; or, to it put another way, the point at which the two filters cross over. For a bi-amp crossover, this frequency is likely to be set to anywhere between 50 Hz and 900 Hz. A middle setting would be around 250 Hz, so all frequencies below 250 Hz would go to the low output, and all frequencies above 250 Hz would go to the high output. A tri-amp crossover splits the frequencies into three different ranges: one for the 'subs' (the subwoofers, for low frequencies), one for the mids (for mid frequencies), and one for the tops (for the high frequencies).

A more modern setup uses a speaker management system to control multiple crossovers and EQ settings.

1.2.4.4 Active loudspeakers

An arrangement suitable for smaller setups is to combine the amplifier and loudspeaker inside the loudspeaker cabinet, simplifying the design and allowing the amplifier to be ideally matched to the loudspeaker. Active loudspeakers are typically less expensive than having a separate amplifier and a passive loudspeaker (equipment tends to be referred to as 'passive' if it does not have a power source, and 'active' if it does). However, active loudspeakers give less flexibility, are more expensive, and are difficult to replace or fix when they break. In Figure 1.7, the sub-frequency loudspeaker is active, and the mid and high loudspeakers are passive.

Sound manipulation techniques of the recording studio

Popular music exists in its current form because of the mechanisms that make it 'popular' – that is, the mechanisms that allow it to be mass consumed. The original mechanism for the mass consumption of music was printed sheet music, leading to the 'Tin Pan Alley' era of music from the 1880s in the USA. Following sheet music, the ability to record and produce music in ever more sophisticated ways, combined with the ability to distribute recorded music, gave us modern popular music (as well as many other types of music). Before these mechanisms, there was no popular music in the way we know it today. Music was passed on in the oral tradition (i.e., folk music), or with expensive musical notation to play what is typically called classical music.

The history of sound recording can be roughly divided into four stages: acoustic, electrical, magnetic, and digital recording technologies.

Acoustic: The earliest sound recording technologies were acoustic horns created in the late 1800s that focused sound energy into a mechanism, which etched a groove into a cylinder or disk. While highly innovative, these setups were particularly limited in frequency range and dynamic range.

Electrical: As technology developed, recordings from 1925 started using microphones and amplifiers connected to an electrical cutting head, significantly improving the quality of the sound signal. The recording medium used essentially the same mechanism of etching grooves into shellac disks.

Magnetic: After the end of the Second World War in 1945, access to German magnetic tape technology allowed a giant leap in sound recording capabilities. Not only was the duration and quality of magnetic tape as a recording medium better than disks, but tape machines also allowed multi-tracking and editing techniques. As magnetic tape replaced shellac disks as a master medium, the new material of vinyl then replaced shellac disks as the primary consumer format from the 1950s.

Digital: Since the 1980s, the ability to encode analogue audio into digital data has resulted in by far the most significant advancements in recording and listening. As a multitrack recording medium, the reliability, fidelity, track count, and

DOI: 10.4324/9781003370406-3

cost-effectiveness of hard drive outboard (followed by computer-based systems) far outstrip those of magnetic tape. As a consumer listening medium, digital audio allowed improvements in fidelity, such as the compact disc. However, in the early 2000s, this fidelity was then swapped for the distribution and convenience that lower-fidelity lossy formats such as the mp3 allow.

Sound sources – recording and electronic sound creation: The traditional music production that takes place in the recording studio can be summarised as a two-stage process of recording sound sources and mixing these sources. As technology developed, alternative electronic sound creation methods became available through synthesisers and samplers. These are explored in Chapter 3, and the associated concepts and tools are discussed in Chapter 4.

Recording studio and electronic music environments – 'recording' and 'building': There is a useful conceptual and practical distinction between the processes of:

- **recording** using (predominantly) microphones and mixing a musical performance in a traditional recording studio, and
- **building** a piece of music using sound creation tools in a software-based 'in the box' environment, referred to here as an 'electronic music environment'.

The tools and techniques of the recording studio are discussed in this chapter. The tools and techniques of the electronic music environment are discussed in Chapter 3.

In keeping with a recurring theme of this book (that categorisation is used for clarity), it is to be noted that in practical use, these different concepts are often blended and juxtaposed for creative effect. The typical production phases of music recorded in the recording studio, electronic music, and live performance events are discussed further in Section 5.3.1.

An essential aspect of this distinction in music production processes is a consideration of distance from the acoustic source material. In an electronic music environment that makes broad use of electronic sound creation techniques, there is a separation from the original recording, as the source material is likely to be either manipulated samples or generated synthesis patches. This separation encourages more extreme sonic manipulation. For example, percussive elements in popular electronic music are likely to be electronic beats, which hardly ever sound like an acoustic drum kit. In rock music, however, there is still the genre convention for the percussive elements to sound more like an actual acoustic drum kit.

Many sound creation techniques for electronic music have merged into the traditional music production process, resulting in an inevitable separation from and reduction in the importance of performance. The requirement for a musician to play accurately and consistently is reduced when so many tools can generate or correct musical parts.

This point is discussed further in Chapter 5, where a case is made for performance that incorporates recording studio and electronic sound creation techniques while remaining an exciting performance with significant exchange of energy.

DAWS: For digital setups (the vast majority), the software used both in the recording studio and in electronic music environments is based on essentially the same type of software: a digital audio workstation (DAW). Some DAWs may be commonly associated with one environment or the other (e.g., Pro Tools for the recording studio and Ableton Live for electronic music). However, the underlying techniques used for music production are the same. The differences in music production across genre and environment – whether rock, hip hop, popular electronic music, or electroacoustic music – are often a factor of how and where these techniques are applied and to what degree.

Mixing: One of the most important frameworks for sound manipulation techniques is the process of mixing. In literal terms, mixing is the process of combining discrete sonic elements effectively. In the traditional model of music production, mixing often occurs as a process discrete from recording. The modern method of mixing has evolved into a highly complicated operation, combining many corrective and creative sound manipulation techniques. The fundamental techniques of mixing are explored in this chapter.

There are as many different approaches to mixing techniques as there are sound engineers and musical projects. However, there are established best practices that can be used to speed up learning and streamline an effective workflow. These best practices are described in the following sections and are accompanied by suggestions for creative and innovative development. It is worth making it clear that there is no single *correct* way to use sound manipulation techniques. Knowing about industry standards and the theory underpinning these techniques is vital to effective workflow. However, this knowledge is also crucial because it can lead to the confidence to experiment effectively and build an innovative personal creative practice.

Processors, effects, and editing overview: Three main areas cover recording studio techniques: processors, effects, and editing. Processors and effects are types of sound manipulation devices typically used in real time. In contrast, the set of audio manipulation techniques employed offline is referred to in this context as 'audio editing'. As discussed in Section 1.2.3, the division between processors and effects as sound manipulation tools is made because, in standard practice, they are used in significantly different ways:

- Processors are sound manipulation tools inserted into the channel to change its sound directly – for example, an equaliser (EQ) that changes the frequency content.
- Effects are sound manipulation tools, such as reverb, that add sonic elements and have their own channel. Proportions of signals from other channels are sent to effects channels so that the effect, or 'wet' channel, can be mixed with the original 'dry' channels.

This signal flow is shown in Figure 2.1.

While the distinction between processors and effects is central to the standard recording studio workflow, it is also worth noting that the umbrella term 'effects processors' can be used as a catch-all descriptor. In many electronic music setups, the term 'effects' covers both effects and processors (e.g., in the Ableton Live documentation at the time of publication). The distinction

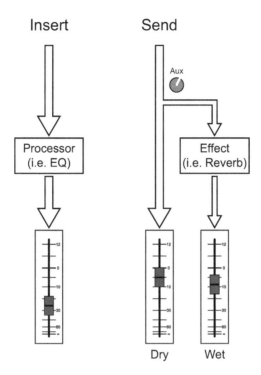

FIGURE 2.1

Insert and send signal flow

between processors and effects is expanded further in the following sections but should not be seen as a hard and fast rule – more good practice for standard recording studio use.

The key concepts and the application of processor, effect, and editing techniques are discussed in the following sections. Specific examples of the processors and effects explored here are typically software plugins, due to the accessibility of computer-based systems such as DAWs and the clearer and more consistent illustration of the techniques that graphical user interfaces are likely to offer. Section 2.4 discusses the importance of stereo image. Section 2.5 concludes with an exploration of techniques for creating audio for live use or use in music production environments – for example, sample packs.

2.1 PROCESSORS

Processors are devices that, when used effectively, give music a clarity, punch, and excitement that would otherwise be impossible. This sound is achieved through an adept combination of many processors, each doing a different job and then ordered and routed carefully, making each musical element work towards a particular production aesthetic.

As mentioned in Section 1.2.3, processors are devices inserted into the channel as, typically, they are used to change the signal's sound directly. In practice, this means that a different processor device is required for each audio channel, which in the days of limited analogue devices was a significantly limiting issue. However, in a DAW, there can be as many instances of a plugin as required, with the main limiting factor being processor load.

There are two main types of processors: dynamic processors and EQs. Chapter 1 discusses processors in the context of live sound issues. Although these are essentially the same devices, in a recording studio, there are typically far more processors in use to sculpt sound in many different ways, from subtly tweaking a sound in small increments to drastically reshaping the sound's character. Because these sound manipulation techniques do not have to take place in real time in relation to the performance, the recording studio engineer has many more options. A studio engineer can take hours reviewing the multiple takes that make up musical parts of a project and edit them into the best possible version for that music. The studio engineer can audition, compare, and experiment with all the processors at their disposal without fear of howling feedback. This time and space separation in the recording studio environment is explored further in contrast to live performance events in Section 5.3.2.

2.1.1 Filters and equalisers

On any standard recording studio analogue mixing desk (or live sound mixing desk) from the last 50 years, the only sound manipulation tool consistently replicated on each channel strip is equalisation. It is hard to over-emphasise the importance of equalisation in mixing.

As explained in Section 1.2.3.1, equalisation consists of a series of filters that adjust the frequency content of a signal. Each EQ is a certain number of bands, which means how many filters it has (i.e., how many different bandwidths of frequency it can affect). A good example of a standard recording studio software EQ, or plugin, is the EQ III – a seven-band EQ that comes with the DAW Pro Tools. A screenshot of the EQ III is shown in Figure 2.2.

Each 'band' of an EQ is a different filter, capable of changing a range of frequencies.

The standard options at the high end of the frequency spectrum are either a high shelf filter or a low pass filter (LPF). Although the names of these filter types may initially appear contradictory, 'high shelf filter' describes the frequencies being changed, whereas 'low pass filter' describes the frequencies that are let through the filter unchanged (or passed). Conversely, at the low end of the frequency spectrum, the standard options are either a low shelf filter or a high pass filter (HPF) (as shown in the top two filters in the EQ III plugin, which can be seen in Figure 2.2). These filter types and their controls are explained in Section 2.1.1.1.

Figure 2.3 shows symbols for the six main filter types: HPF, LPF, high shelf filter, low shelf filter, peak filter, and notch filter. Most devices that allow multiple types of filters use these symbols or similar versions of them.

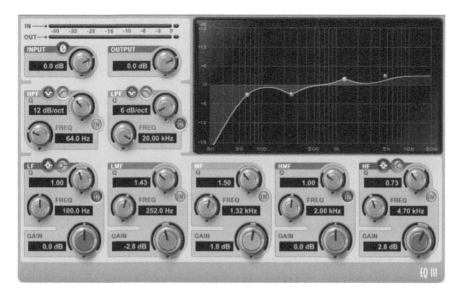

FIGURE 2.2
Pro Tools EQ III plugin showing a range of filters in use

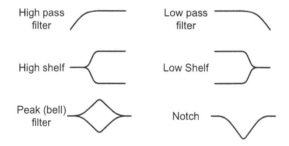

FIGURE 2.3
Symbols for the six main filter types

2.1.1.1 Filter types and controls

2.1.1.1.1 High pass filters and low pass filters

HPFs let frequencies through, or pass, over a particular frequency, called the cutoff frequency, which is measured in hertz (Hz). Frequencies below the cutoff are reduced until they disappear. High pass filters are typically used to remove a large amount of low frequencies – for example, a high pass filter at 100 Hz to remove bass spill in a vocal track.

High pass filters have two controls: the cutoff frequency and the filter's slope. For a high pass filter, the slope sets how sharply the filter removes frequencies under the cutoff frequency and is measured in decibels (dB) per octave. The technical definition of the cutoff is the point at which the filter has already reduced the frequency content by 3 dB. A gentle slope of -6 dB per octave (dB/oct) or -12 dB/oct will let more low frequencies through under the cutoff and is likely to sound more natural than a more aggressive slope such as -24 dB/oct. Figure 2.2 shows a high pass filter with a cutoff of 64 Hz and a slope of -12dB.

LPFs let frequencies through, or pass, under the cutoff frequency. Accordingly, they are typically used to remove a large amount of high frequencies over the cutoff. As with a high pass filter, the low pass filter also has cutoff frequency and slope controls. For a low pass filter, the slope sets how sharply the filter removes frequencies over the cutoff frequency. In the opposite way to a high pass filter, a low pass filter with a gentler slope will let more high frequencies through. As with a high pass filter, the slope of the low pass filter will determine how natural or aggressive it sounds. Low pass filters are essential for synthesis and are discussed further in Section 3.3.

2.1.1.1.2 High shelf filters and low shelf filters
High shelf filters boost or cut all frequencies above the cutoff frequency by a certain amount, called the gain, measured in dB. High shelf filters have three controls: the cutoff frequency, the gain, and the Q. The Q is named after the quality factor of a filter and – similarly to the slope of a HPF or LPF – describes how steeply the filter changes. Q is not measured in dB/oct; it is a number that represents the ratio between the centre frequency and the bandwidth (see Section 2.1.1.1.3 for descriptions of these terms). The higher the Q, the steeper or more aggressive the filter. As with the slope, a more aggressive Q (e.g., 4 or more) is likely to sound less natural and more likely to produce unwanted distortion. Figure 2.2 shows a high shelf filter with a cutoff of 4.7 kilohertz (kHz), a gain of 2.8 dB, and a gentle Q of 0.73.

Low shelf filters work in the same way as high shelf filters, except that they boost or cut all frequencies below the cutoff frequency.

2.1.1.1.3 Peak filters and notch filters
Peak filters (also known as bell filters) either boost or reduce frequencies around a specific frequency, called the centre frequency, with the amount of boost being a positive gain in dB. Notch filters (or band stop filters) cut frequencies around the centre frequency at maximum attenuation (i.e., they have no gain).

The bandwidth is the amount of frequencies on either side of the centre frequency affected by either a peak or notch filter, and is determined by the Q value. Somewhat counterintuitively, the higher the Q, the fewer frequencies affected either side of the centre frequency, because Q is the ratio of the centre frequency to the bandwidth. For example, a peak filter with a centre frequency of 500 Hz that starts boosting at 400 Hz and stops at 600 Hz has a bandwidth of 200 Hz (600 - 400), so its Q is 500 divided by 200, which is 2.5. If the bandwidth narrows to 100 (so the filter starts at 450 Hz and stops at 550 Hz), then the Q is now 5.

As with the slope value of HPFs and LPFs, and as with the Q of shelving filters, a more aggressive Q (e.g., 4 or more) is likely to sound less natural and more likely to produce unwanted distortion.

Figure 2.2 shows two peak filters: one reducing and one boosting. The reducing peak filter has a centre frequency of 252 Hz, a gain of -2.8 dB, and a Q of 1.43. The boosting peak filter has a centre frequency of 1.32 kHz, a gain of 1.5 dB, and a Q of 1.5.

2.1.1.2 Equalisation in use

Unless the devices used are linear phase equalisation devices, which typically have high latency and processor load, standard equalisation uses phase shifting to alter frequency content. Because this phase shifting introduces a small amount of distortion, it is generally best to reduce unwanted frequencies (and then increase the gain if needed) rather than boosting the frequencies that need to be more prominent. This way, the more prominent frequencies are not phase shifted or altered tonally. As well as this, boosting frequencies tends to immediately sound more pleasing simply because they are louder, which may obscure a more critical assessment of the sound and how it fits with an overall production aesthetic.

If the aim is to maintain the tonal purity of the original sound, it is always worth considering whether this can be achieved by getting a better sound at source. Good sound at source is a fundamental pre-production issue for either tracking in the recording studio or playing in live events. For example, are the microphone choice and technique optimal? If the signal needs to be brighter, could the microphone be placed 'on axis' (i.e., directly at the source)? Could another, brighter microphone be tried? Could the musician consider a different tone in their instrument to suit the performance and reduce the risk of a higher noise floor?

2.1.1.3 Transparency and character

As mentioned previously in Sections 1.2.3.2.1 and 2.1.1.2, a corrective approach to sound manipulation results in a more effective sound but with no significant change to the character of the sound. Some processors are particularly suited to corrective methods, as they sound more transparent. On the other hand, other processors have a characteristic sound that alters the signal's tone in a particular way, so they are well suited to more creative approaches. The standard plugins that come with a DAW are likely to behave more transparently – for example, the EQ III plugin in Pro Tools, shown in Figure 2.2 – so they are well suited to corrective equalisation approaches.

EQs with a characteristic sound or 'colour' tend to be modelled on previous analogue hardware models. Due to early EQs' non-linear analogue operation, they impart a particular characteristic to the signal – for example, the Pultec EQP-1A outboard (the universal audio plugin is shown in Figure 2.4) and EQs from classic Neve and SSL mixing desks. Each of these EQs can add sonic colouration most simply described using subjective terms – for example, 'grit', 'brilliance', or 'smoothness'. Many plugin versions of these devices have been created using software models of the analogue circuitry by companies such as Sonimus, Universal Audio, and Waves. Devices that colour the tone tend to have fewer controls, offering a more limited but tried-and-tested method of operation.

FIGURE 2.4
Universal Audio Pultec EQP-1A EQ plugin

FIGURE 2.5
Variety of Sound Tokyo Dawn Labs Slick EQ plugin

Not many plugins offer coloured equalisation of different types, but an exception is the SlickEQ – a free EQ plugin from Tokyo Dawn Records shown in Figure 2.5. Instead of a Q control, the SlickEQ offers four EQ-type curves – labelled as British, American, Russian, and German – each with a different sound, or frequency response. Harmonic distortion (saturation) is introduced into the filter network with the EQ SAT button. While not a typical method of equalising with colour, the SlickEQ offers a helpful demonstration of the different types of colourations available.

A valuable technique for hearing the colouration that any processor adds to the signal (either EQ, compressor, or limiter) is to solo the channel and push the change far more than would be necessary. This change can then be lowered to a more subtle level, with a greater awareness of what that device is doing to the signal's sound. Once familiar with the character and general response of the device, it is generally considered best practice to 'mix in context' – that is, not to use soloing when applying sound manipulation techniques.

2.1.2 Compressors

One of the most significant trends in music production over the last 50 years is that recorded music has been made to sound ever more exciting and bigger. As well as EQs, the other processors crucial to create aural excitement and size are the devices that allow dynamic range compression: compressors, limiters, and maximisers. Of the three, compressors are the most widely used and are focused on for most of this section. Limiters and maximisers are discussed in Section 2.1.2.6.

The dynamic range of a signal is the difference between the quietest and the loudest elements. Compression reduces this difference by reducing the peaks of a signal by a certain ratio. For example, if a compressor (with a hard knee – see Section 2.1.2.1) has a ratio of 2:1, for every two dB over the threshold, only one dB is output. As these peaks are reduced, the signal's gain is raised (by a control typically marked 'makeup gain'), resulting in a louder-sounding overall signal. (Technically, this is called downward compression, as it reduces high amplitudes; the other type of compression is upward compression, which raises low amplitudes. Upward compression is not generally useful for music production, so it is not discussed here. (For further information, a great explanation of upward and downward compression and expansion can be found in the Ableton Live Audio Effect Reference in the 'Dynamics Processing Theory' section, which at the time of writing can be found online.)

In the context of commercially released recordings, many different compressors (and many different EQs) are likely to be used. There are three levels of use for compression: at the individual channel level (e.g., a snare drum), at the group/bus level (e.g., a drum group/bus), and at the overall mix level (e.g., when mastering the whole piece of music). Each of these requires different compression techniques and a clear understanding of what the controls of a compressor do to the sound at either channel, group/bus, or mix level. These techniques are discussed in the following sections covering best practices in the recording studio. The application of these techniques in an on-stage live performance event context is discussed in Section 5.3.

2.1.2.1 Controls

As discussed in Section 1.2.3.2.2, the main controls of a compressor are the threshold and ratio.

Threshold and ratio: The threshold is the point at which the compressor starts reducing the level, expressed in dB. The lower the threshold, the sooner the compressor will act and the more the signal will be compressed. The ratio is how much the level is reduced, expressed as a ratio. An example of a mild compression ratio would be 2:1, and a heavier ratio would be 8:1 or higher.

The speed at which the compressor reacts also makes a significant difference to the sound of the signal. The compressor's speed is determined by the attack and release controls, the type of compressor, and its response characteristics.

Amplitude envelopes and the recording studio: The attack and release controls for a compressor set how quickly it starts to compress and stops compressing. They are time-based controls that react to the signal's amplitude envelope. A signal's amplitude envelope is how the signal's volume changes over time. A single instrument or vocal's amplitude is likely to have a characteristic rise and fall, which will tend to repeat.

A musical amplitude envelope can be summarised into three sections: attack, sustain, and release. The attack section is the time taken to go from no level to maximum level (for that note or hit), the sustain is the section that stays at a steady level, and the release is the time taken to die away.

The attack and release controls in a compressor (or any time-based control in any processor) are not the same as the attack and release sections of a signal's amplitude envelope. However, they may affect these signal transients, as described in the following sections and Section 2.1.2.3.

It is also worth noting that envelopes are a vital part of sound synthesis and are somewhat different in that context. Envelopes in the context of sound synthesis are discussed in Section 3.4.1.

Attack: The attack sets how quickly the compressor reduces the signal and is measured in milliseconds (ms). Counterintuitively, a slower attack can allow an initial burst of the signal's attack transient to come through before the signal is compressed, which can accentuate the initial crack of a snare drum, for example. A compressor with an attack that is too fast can often sound like an initial sharp transient is choked, but a faster attack will give greater overall control of the peak signal levels.

As a very general rule, for less percussive sounds such as vocals, a shorter and more controlling attack as low as 5 ms is not likely to adversely affect the signal. For a snare drum, a setting between 16 and 25 ms for a fast voltage controlled amplifier (VCA) or digital type compressor is likely to allow enough transients through to give the sound an initial 'crack' and is typically used to accentuate this initial transient while making the whole snare hit sound larger.

Attack and release times depend heavily on the compressor used, especially if the compressor is either hardware or modelled on a hardware device. Ideally, the attack and release are used to fine-tune how much the attack and release transients are accentuated.

The effect of the different compressors on their sound is summarised in Section 2.1.2.2.

Release: The release sets how quickly the compressor resets, so a fast release setting will make a compressor reset quickly, which can raise the sustain and release section of a note or hit. (The amplitude envelope for a note or hit can be summarised into the attack, sustain, and release sections – these are explained in Section 3.4.1.) A longer release time will keep the signal compressed, holding the sustain and release sections at a low level. However, for quick successive notes or hits, this may mean that the compressor is still compressing and may reduce some initial transients.

Many compressors will have an 'auto' setting on the release to automatically vary the release dependent on the audio signal (sometimes referred to as 'program dependent'). Auto is to allow for more complex signals, typically made up of more than one instrument, so this setting is generally used for compression applied to a group/bus or mix. Complex signals are likely to have sounds with different sustain and release times, so automatic release will cause the compressor to adjust automatically to avoid the unwanted pumping that fast compression can cause in complex signals. (Pumping and its creative use are discussed in Section 2.1.2.4.)

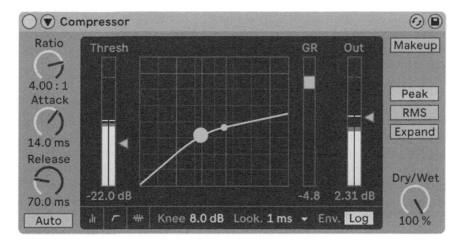

FIGURE 2.6
Ableton Live compressor plugin in transfer curve mode

Knee: The knee of a compressor is how far below the threshold the compressor will start to compress, expressed in dB. A knee of 0 dB is called a hard knee, as the compressor will immediately compress everything over the threshold by the amount set by the ratio. The higher the knee setting, the softer the knee of the compressor, which will impart a softer sound as the compressor starts to work more gradually. Figure 2.6 shows the Ableton Live compressor in transfer curve (also called transfer function) mode. This curve shows the output in the vertical or 'y' axis plotted against each input in the horizontal or 'x' axis. The large yellow circle shows the threshold and the smaller yellow circle shows current signal output.

The transfer curve for a compressor starts at the bottom left, raising as a straight diagonal as every 1 dB of input is 1 dB of output (i.e., no compression). As the compressor starts to reduce the level signal, the angle of the line reduces, showing how much compression will be applied at those levels (after the attack time). In Figure 2.6, the transfer curve is straight until it reaches the amount below the threshold set by the knee. As the threshold is -22 dB and the knee is 8 dB, the compressor starts to compress gradually when the input signal is -30 dB and only reaches the ratio of 4:1 when the input signal reaches 8 dB over the threshold (-14 dB).

2.1.2.2 Types of compressors and character

As discussed in Section 2.1.1.3, the plugins that come with a DAW (e.g., the Ableton Live compressor shown in Figure 2.6) are likely to have a transparent sound – that is, they will not tend to change or colour the sound of the signal.

As with EQs, compressors with a characteristic sound, or 'colour', tend to be modelled on previous analogue hardware models. Due to the analogue nature of the early compressors, they impart a characteristic sound. The most notable hardware compressors are those that impart a particularly desirable sound due to their specific non-linear operation. The methods of operation of these hardware compressors can be summarised into four categories: vacuum tube, optical, field effect transistor (FET), and VCA.

Vacuum tube and variable-MU compressors: The first compressors were developed to control audio levels in radio transmission in the 1930s, using vacuum tube (or valve) amplification. The amount a vacuum tube amplifies the signal is called its amplification factor, or MU. In normal amplifier operation, the MU is as linear as possible – that is, it amplifies large signals by the same amount as small signals. In the 1950s, vacuum tubes were produced with a non-linear or variable MU. In practice, this means that when a signal in a variable MU compressor exceeds the threshold, the vacuum tube does not amplify as much; and the more the signal increases, the *less* the signal is amplified. The ratio of a variable MU compressor is determined by the signal strength – which, combined with the inherent non-linear nature of vacuum tubes, gives variable MU compressors a soft-sounding, often full and warm compression characteristic.

The earliest device to employ variable MU vacuum tubes for music production was the Fairchild 660 (and the two-channel 670), produced in the 1950s and used heavily on a wide range of music, from the Beatles to Motown and Pink Floyd. Figure 2.7 shows a screenshot of the Avid plugin version of the Fairchild 660. Another noteworthy example of a classic variable MU compressor is the Manley variable MU limiter compressor.

Optical compressors: In the late 1960s, an alternate system to vacuum tube compression was released using the combination of an electronic light and a light-dependent resistor. As the signal strength increases, the light shines brighter, and the light-dependent resistor reduces the amplification factor. Due to the time it takes for the light to be recognised, optical compressors tend to have relatively slow attack times. The non-linear nature of the light and light-dependent resistor account for particularly non-linear releases, starting fast but slowing down as the release continues. Consequently, optical compressors tend to impart a gentle and

FIGURE 2.7
Avid Fairchild 660 plugin

smooth sound, letting quick transients through. The Teletronix LA2A and LA3A levelling amplifiers are the earliest and most noteworthy compressors. A screen-shot of the Universal Audio plugin version of the LA-2A is shown in Figure 2.8.

FET compression: One of the most famous and characterful hardware compressors is the UREI 1176, also released in the late 1960s, which makes use of FETs. The 1176 has a particularly aggressive tone – partly due to the characteristic bite and grit of FET distortion, but also because FET compression is capable of much faster compression than vacuum tube or optical compression. A screenshot of the Universal Audio plugin version of the 1176 is shown in Figure 2.9.

VCA compression: A VCA is the most modern and common type of analogue compressor. In hardware, a VCA is an integrated circuit (a chip) that allows any kind of attack and release curve possible. The most notable 'vintage' VCA compressors are the dba 160 and the bus compressor from the SSL 4000G mixing desk, often modelled as a compressor to 'glue' a bus/group or mix together (the term 'glue' is explained below). These are fast and bright compressors, giving the popular music of the 1980s much of its characteristic sound. Similarly, the Focusrite Red 3 compressor created in the 1990s is another notable hardware VCA compressor with a particularly punchy sound, often used for bus compression. Standard modern digital compressor plugins are often based around the VCA design and are the most transparent type of compressor, ideal for a cleaner sounding level or transient control.

'Glue': The term 'glue' is regularly used in mixing with compression and refers to the quality of a compressor to bring together the different elements of a bus/group or mix. A music production environment, where many musical elements

FIGURE 2.8
Universal Audio LA-2A plugin

FIGURE 2.9
Universal Audio 1176 plugin

are intentionally isolated to process more cleanly, can result in a lack of musical cohesion, especially on drum kit mixes. A suitable compressor inserted into a bus or group can help to restore musical cohesion or glue it together. A compressor that is good at glueing these elements together will typically occupy the middle ground between lots of character and complete transparency, but is likely to be able to have a fast attack and a non-linear release. The most obvious example of a compressor well suited to glueing a mix together is the SSL 4000G VCA bus compressor, mentioned in the previous section – which, not very incidentally, is the compressor used to model the glue compressor in Ableton Live.

2.1.2.3 Level evening and transient control

There are two main uses for a compressor in the recording studio mixing process: level evening and transient control.

As described in Section 1.2.3.2.2, the primary use of a compressor in live sound is to even out the dynamic range of a signal – that is, to make the ongoing signal level more consistent. Level evening is also one of the core uses of a compressor in the recording studio mix process. For a compressor to do this, the attack must be fast enough to not let through sharp transients, and the release must react slowly enough not to accentuate sustain and release sections.

The other core use of a compressor is to use the attack and release characteristics of the compressor to bring out certain aspects of the signal's transients – for example, the raising of the attack and the lowering of the release to accentuate the sustain and release section of the signal's amplitude envelope, as discussed in Section 2.1.2.1. The settings and character of a compressor can dramatically alter a signal's transients in a very prescriptive, threshold-dependent way. Accordingly, compressors are particularly useful for mixing, as they allow consistent transient control, in contrast to a transient shaper. A transient shaper reacts to signal *differences* rather than at a specific threshold, so that it will alter quieter signals as well as louder ones. Transient shapers are discussed in Section 2.1.3.

Much like many other creative distinctions discussed in this book, these two uses of compressors are not at all a binary separation. Any compressor used for transient control may also level out a signal, and any compressor used for levelling out a signal may also change the transients. The distinction is offered so that we can be critically aware of the point of a compressor in the signal chain. Some compressors may be better suited to level evening and some to transient control, and their performance is also likely to differ in the context of the signal type and mix situation.

2.1.2.4 Side-chaining

By default, a compressor is triggered by its track's signal. An alternate use of a compressor is a side-chain (or key input) – that is, an alternate input. The side-chain can be a filtered version of the track's signal or a completely different signal.

A filtered version of the signal used as an input is particularly useful for reducing signal levels when only specific frequencies are prominent. For example, the low frequencies of a wideband signal may be causing the compressor to overly compress when those frequencies are present, so a high pass filter can be used to take them out of the signal triggering the compressor. Suppose that the signal triggering the compressor is limited to the frequencies that make up a vocal 's' sound (4–7 kHz). In that case, the compressor will only compress when those signals are present and can reduce sibilance – a basic example of 'de-essing'.

A typical modern application of a different signal as a side-chain is using a kick drum as the input for a bass guitar (or any other musical element), so that whenever the kick drum plays, the level of the bass guitar is reduced (or 'ducked'). There is then a space for the kick to punch through. If used more dramatically, side-chaining can change the character of the music, giving a pumping sound used in a wide range of popular music that the developing sounds of popular electronic music, RnB, and hip hop have influenced.

2.1.2.5 Multiband compressors

The compressors discussed in the previous sections are wideband, so they affect all signal frequencies equally. Multiband compression splits the signal into several frequency bands, each with its own compressor. Multiband compression is beneficial for altering the dynamics of different frequency components of complex sounds and is traditionally used for mastering. The increase in computer processing power and plugin availability has meant that multiband compression can be more freely used on individual tracks, typically giving a more hyped sound when used aggressively. For example, take a bass synthesiser patch with less compression in the middle and treble frequency bands than the bass frequencies. In that case, the sound can be made both brighter (as the mid and high frequencies may have more dynamic range and impact) and thicker (as the bass frequencies may be evened out).

2.1.2.6 Limiting

Limiting is a type of dynamic range compression with a very high ratio, typically fixed around 15:1 to infinity (∞):1. A limiter gives very aggressive dynamic range compression when a signal reaches the ceiling value (at just below 0 dBFS if the limiter is in place for mastering or to guard against clipping). As the name suggests, a limiter does just that: it limits the signal from increasing.

Because of this, limiters tend to have fewer controls than compressors, ranging from just a single ceiling value to more flexible limiters with other controls, such as:

- gain, to raise the signal closer to the ceiling
- threshold, which on many mastering limiters acts as an inverse gain control – that is, when turned down, it raises the signal closer to the ceiling
- release, which – as with a compressor – sets the time taken to stop limiting after the signal drops below the ceiling.

Some limiters also have knee (see Section 2.1.2.1) and attack or look ahead controls, allowing them to be more forgiving to attack transients.

Traditionally, limiters were only used as a part of the mastering process. However, as with multi-band compression, as modern popular music has become more compressed and the means to limit more accessible, the more aggressive character of limiters (compared to compressors) is increasingly used to create a hyped, popular electronic music-like characteristic.

The other type of limiting, reserved almost exclusively for mastering, is maximising: a kind of digital peak limiting that uses digital analysis of the source signal to maximise the loudness of its output. Maximisers can create mixes far louder than a conventional limiter and are the devices driving the trend of ever-louder music masters – often called 'the loudness wars'.

2.1.2.7 Parallel compression

As discussed in the introduction to this chapter, in regular operation, processors are devices inserted into the channel to remove or replace audio elements.

However, sometimes compressors can be driven to create a characterful dense sound. If inserted, this sound is likely to be too much on its own. However, if that sound is blended in with the original, it can be particularly effective, as the original dry signal lends dynamic range and the wet signal adds the characterful sound of the driven compressor.

Some compressor and limiter plugins have a wet/dry control, allowing this blending to happen when used as an insert; otherwise, a proportion of the signal will need to be sent to the compressor and limiter on a separate aux, return, or effect channel.

Parallel compression is essentially making use of a processor as an effect. While it is rare to see EQs used as effects (as this is likely to cause phase problems), parallel compression (also sometimes referred to as New York compression) is often used to introduce some dynamic and sonic variety to a mix.

2.1.3 Transient shapers

Transient shapers change the attack and sustain sections of a sound. A transient shaper reacts to signal differences rather than at a specific threshold (as compressors do, see Section 2.1.2.3), so a transient shaper alters quiet and louder signals (rather than just louder signals, as compressors do). Typically, the controls of a transient shaper are an attack control and a sustain control that can either boost or cut that element of the sound's envelope by a certain amount measured in dB.

2.1.4 Gates and expanders

2.1.4.1 Gates

As explained in Section 2.2.3.2.2, a gate (or noise gate) only lets through sound that has a higher signal level than the threshold, so that unwanted noise is removed or reduced. Figure 2.10 shows the Pro Tools expander/gate plugin operating as a gate. The transfer function in the upper part of the figure indicates that any signal under -18 dB will be reduced by the amount set by the range control – -35 dB in this case. Figure 2.11 shows the Ableton Live gate device that uses a graph to display how the gate reacts over time (rather than a transfer function).

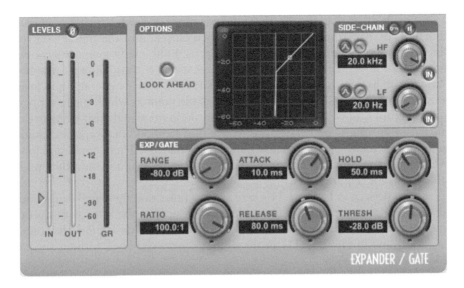

FIGURE 2.10
Pro Tools Expander/Gate plugin operating as a gate and showing transfer function

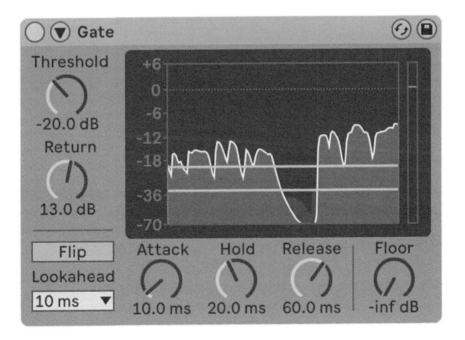

FIGURE 2.11
Ableton Live Gate plugin showing signal over time

Controls other than the threshold include **range** or **floor**, which is how much reduction the gate applies to the signal when closed, and the timing controls, which are: **attack** – how quickly the gate opens; **hold** – how long it stays open; and **release** – how quickly it closes. Some gates – such as those shown – also have a **return** (or **hysteresis**) control, which separates a gate's opening and closing thresholds. With a return set, the threshold is the point at which the gate opens, and the return is how many dB below the threshold the signal must go before the gate closes. A return control helps to reduce the likelihood that rapidly changing signals make a gate open and close quickly, or 'chatter'. These upper and lower levels to the threshold can be seen in the time response graph in the Ableton Live gate screenshot (Figure 2.11).

As can be seen in Figure 2.11, there are five times that the signal drops under the threshold, but the signal only has its gain reduced in the fifth, as this is only time that the signal drops below the return level. The dark grey area (under the return line in Figure 2.11) in the display shows the level without the gain reduction.

A more flexible and clinical alternative to gating is the silencing tools discussed in Section 2.3.1.2 (e.g., strip silence in Pro Tools).

The benefit of using a gate over silencing tools is that if the settings remain the same, the noise reduction will be consistent across the track. Also, because silencing tools are editing tools, they are of minimal use in a live environment.

2.1.4.2 Expanders

Expanders are like gates, as they let sound through over a particular threshold. However, for signals below the threshold, expanders reduce more gently, by a ratio, rather than immediately dropping to the **range** or **floor** value, as a gate does. (Technically, as with compression, this is called downward expansion, as it reduces *low* amplitudes. The other type of expansion is upward compression, which raises high amplitudes. Upward expansion, like upward compression, is not generally useful for music production, so it is not discussed here.)

Expanders sound like a softer, less aggressive gate, as the signal somewhat fades in. The 2:1 ratio indicates that for every 1 dB under the threshold, the signal is 2 dB quieter. This gentler slope can be seen in the transfer function slope in Figure 2.12 and results in an expansion of the dynamic range, essentially making the quiet signals quieter.

2.1.4.3 Corrective and creative use of gates and expanders

As a corrective audio manipulation technique, gating and expanding in the recording studio is a more traditional live approach to cleaning up a signal. Compared to the offline editing techniques or silencing tools discussed in Section 2.3, which give far more control and editability over the removed audio, gating and expanding are quite limited.

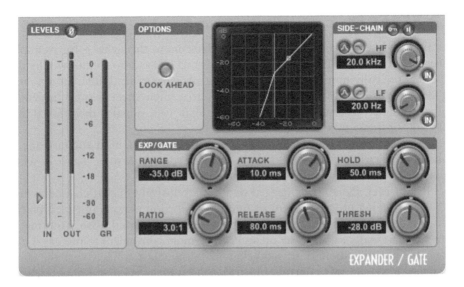

FIGURE 2.12
Pro Tools Expander/Gate plugin operating as an expander

However, as well as corrective tools, gates and expanders can give dramatic percussive effects. A typical example is side-chaining a gate from another input, providing a more extreme version of side-chain compression.

2.1.5 Fader riding and automation

Fader riding refers to the continuous adjustment of the main volume fader of a channel (typically the main vocal) to make the sound fit better in the mix. Fader riding allows a more fluid and creative approach to dynamic adjustment than just compression.

As discussed in Section 1.2.3.2.2, in the context of live sound, it is difficult to get a consistently accurate balance in a live performance event. In the recording studio, however, the performances are offline, so they can be listened to repeatedly. In a traditional analogue recording studio setup, listening to the recording and then practising riding the fader is one of the core tasks of a mix engineer. As digital mixing desks and DAWs become more widely used, fader automation has made it possible to prescribe the changes to the volume of the channel (as well as many other parameters of a DAW's mixer, processors, and effects). Automation is displayed in DAWs as a graph showing parameter changes over time.

It is, of course, impossible to replicate the offline process of creating automation graphs for a live performance event. However, there are fader-riding plugins – most notably Vocal Rider by Waves, which does a good job of a standard fader ride, essentially acting like a gentle and highly program-dependent compressor. Using fader riding plugins in live performance is discussed in Section 5.3.3.6.

2.2 EFFECTS

As discussed in Section 1.2.3, effects add elements to the signal – for example, reverb, delay, or distortion. Rather than inserting effects into channels individually, it is typically more practical (and more efficient) to send different amounts of 'dry' signals from several other channels into a single effect channel (generally called auxiliary, return, or effect channels). That 'wet' signal is then mixed into the front-of-house mix.

Using processors as inserts and effects as sends is not a hard and fast rule. It is more of a starting point for good practice and something to be aware of when applying more experimental approaches with these tools.

If, however, the effect is needed to apply exclusively to that channel where it goes through a processor – for example, a distortion plugin (an effect) placed before a compressor (a processor) – it will need to be inserted. Also, inserting an effect is marginally easier to set up and can be more convenient to mix, as it will not add an extra auxiliary or send channel (although it is likely that the effect will need a wet/dry mix control of its own so that the effect does not overwhelm the sound).

It is also worth bearing in mind that system reliability is the most critical factor in system setup, especially for live performance events. Awareness of the most effective use of processing power (e.g., dictated by plugin count, track count, or the need to record) should be considered to avoid glitches or a system crash.

Because effects add elements to the signal, they are discussed in terms of the different types of character that they offer.

2.2.1 Reverb

Reverberation (or 'reverb', as it is more commonly known) is the name given to the reflections of sound from the interior surfaces of an enclosed space, making that sound continue to be heard for a short period. In the recording studio, sounds are often captured with direct microphone techniques to capture a strong signal with minimal natural reverb and to reduce spill from other instruments. This direct signal then allows the flexibility of adding different types of highly controllable artificial reverb as part of the mixing process (as well as allowing deeper use of processors).

In the early days of recording studio history, reverb was added by playing back recorded parts in suitable sounding enclosed spaces (called echo chambers, despite rarely creating distinct echoes) and re-recording that sound. More recently, reverb is typically added artificially using either a reverb plugin or outboard devices; and as with EQs and compressors, reverb devices tend to use broadly similar controls.

2.2.1.1 Reverb controls

To simulate desirable sounding reverb, either a simulation of an enclosed space type (e.g., rooms or halls) or a simulation of mechanism types (e.g., plate or string) is typically used. The selector for the reverb type tends to form the primary control for a digital reverb effect. These types are expanded on and explained in Section 2.2.1.2. A good example of a plugin reverb illustrating buttons for these types is D-Verb in Pro Tools, shown in Figure 2.13.

The other standard controls are the time/decay and the pre-delay. The time/decay controls the time it takes for the reverb to tail off. Generally, reverb times shorter than about two seconds give a clearer sound but lack richness, while reverb times longer than two seconds are richer but lack clarity. The pre-delay sets an amount of delay in milliseconds before the reverb starts – which, because it makes the reverb sound further away, makes the source of the sound (the dry signal) appear closer to the listener, as well as potentially making the reverb space sound bigger.

Other standard reverb parameters are size, early reflections, reverb tail, and damping. The size control makes the space or mechanism of the reverb sound larger or smaller. Early reflections are the individual reflections off the walls of an enclosed space that typically happen in the first 100 ms and give spatial information. The reverb tail (or sometimes just the reverberation, as opposed to the early reflections) is the ringing out of sounds that cannot be distinguished from each other and keep sounding out after 100 ms, giving the reverb its rich quality. Often reverb devices will allow control over the timing and amount of both the early reflections and reverb tail. Dampening is the high frequency loss as reverb continues due to the sound hitting more absorbent materials. If damping is increased in a reverb effect, the high frequencies will be reduced over time, and the reverb will sound warmer.

FIGURE 2.13
Pro Tools D-Verb

2.2.1.2 Types

As described in the previous section, reverb effects are separated types characterised by different sorts of spaces, such as rooms or halls, or mechanisms, such as plates or springs. In reverb plugins, the types in the following sections are algorithms or impulse responses that aim to replicate the original spaces of mechanisms.

2.2.1.2.1 Rooms and halls

Room and hall settings replicate how sound reverberates naturally in an enclosed space. Typically, room settings simulate a smaller space, with strong and less complex early reflections combined with a dense and short reverb tail. Conversely, hall settings simulate a larger space, with early reflections that build slowly, combined with a longer and more diffused reverb tail. There are several other descriptors for spaces, such as churches, for very large and highly diffused spaces, and chambers, which simulate the echo chambers mentioned in Section 2.2.1 and typically have very few early reflections.

2.2.1.2.2 Plate and spring reverb

Shortly after echo chambers, reverb was simulated by connecting a contact loudspeaker to the surface of a material, such as a large plate of metal or a spring, so that the sound waves would reflect through the material. Contact microphones would then be placed to pick up these reflections and allow them to be mixed into the dry recordings. Because of the mechanical nature of these devices, this reverb was not particularly natural sounding; however, in some circumstances, it was found to be particularly useful.

Plate reverb tends to sound unnaturally bright and dense, but this can be very useful for making percussive signals sound piercing and more powerful.

Spring reverb tends to sound twangy and not like an enclosed space. However, it gives a rich and colourful character to the electric guitar, primarily providing the 'surf guitar' sound of the 1960s. A mechanical spring reverb unit is still a common addition to many guitar amplifiers.

2.2.1.2.3 Algorithmic and convolution reverb

In digital reverb effects, reverb is created using either algorithmic or convolution methods. Algorithmic reverbs simulate the early reflections and reverb tail of reverb using computer algorithms – that is, instruction sets – and are significantly more computationally efficient than convolution methods.

Convolution reverb uses a sample of an acoustic space, called an impulse response, and then uses the mathematical convolution process to create the reverb from the incoming signal. Because of this process, convolution can produce far more natural-sounding reverb, but this takes much more computer processing power. Figure 2.14 shows a screenshot of the Ableton Live (Max for Live) device convolution reverb.

As a general rule, if a reverb is not described as a convolution reverb (not necessarily in the title), it is likely to be an algorithmic reverb.

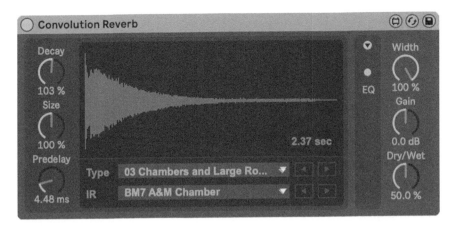

FIGURE 2.14

Ableton Live Convolution Reverb Max for Live plugin

2.2.2 Delay

Delay, in its simplest form, is an effect that delays the signal by several milliseconds or seconds and mixes the delayed signal with the original signal. If the delay is less than approximately 50 ms, combining the delayed and original signal creates a thickening and blurring of the sound. If the delay is more than approximately 50 ms, combining the delayed and original signal creates a distinct duplicated echo of the sound. 50 ms should be taken only as a rough guideline, as the time it takes for a delayed copy of a sound to be heard as a distinct sound will depend on several factors – the most important being the attack.

2.2.2.1 Delay controls

The complexity of a delay effect will depend on the number of delays it allows and the manipulation that can be applied to these delays.

Figure 2.15 shows a screenshot of the Ableton Live Filter Delay device. Filter Delay has three delay lines: one for the left input channel, one for the left and right input channel combined and one for the right input channel. Each of these can be filtered independently by linked HPFs and LPFs shown in the small graphs. The most common controls for a delay device are the delay time, the feedback, and the dry amount or wet and dry mix.

The delay time is measured in milliseconds or a rhythmic amount relative to the tempo of the DAW. This rhythmic amount is typically referred to as sync mode and is displayed as a rhythmic musical value. In Figure 2.15, the left and right (L+R) delay is set to a very short 10 ms; whereas the left delay is set to Sync. Sync allows the delay to be set to several 16th notes – in this case, 4 × 1/16th notes, otherwise known as a quarter-note, one beat, or a crotchet.

The feedback control sets how much of the delayed signal returns to the input of the delay line, causing repetitions of the delay. If the feedback is set very

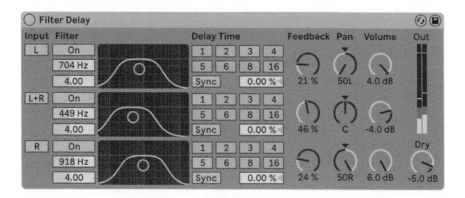

FIGURE 2.15
Ableton Live Filter Delay plugin

high, then the delayed signals can combine to get ever louder and louder until they become noise.

The dry amount (as shown in Figure 2.15), or wet and dry mix, sets how much of the original (dry) signal is output by the device, so in regular operation on a wet send channel, this will be set to minus infinity dB or 0%.

2.2.3 Modulation effects

The category of modulation effects is a broad, diverse set of effects, each with its own sound, but all making use of modulation in some way. Modulation is discussed further in Chapter 3 in the context of synthesis and sampling. It is introduced here in the context of recording studio effects as the underpinning concepts are the same.

'Modulation' in the context of audio manipulation is a term taken from telecommunications technology where one signal, called the carrier, has one or more of its properties (e.g., volume or frequency) varied by another signal, called the modulator. In audio manipulation, this means that a property of the musical signal (the carrier) is altered by a control signal (the modulator). In the context of recording studio effects, the modulator is often a low frequency oscillator (LFO). An LFO creates a repetitive control signal (e.g., a sine wave), which has a frequency that is typically lower than 30 Hz. The parameter of the carrier being modulated largely determines the effect type. For example, tremolo uses an LFO to modulate the carrier's amplitude (i.e., volume). Figure 2.16 shows a diagram representing the modulation that causes tremolo.

2.2.3.1 Flange and chorus

Flange and chorus effects work by duplicating the original signal, delaying that duplicate signal, and then modulating the delay time of the carrier before being mixed back into the original signal. Flange typically uses shorter delay times than chorus – typically between less than 1 ms and up to 5 ms. This short delay time

FIGURE 2.16
Modulation causing tremolo

creates deep frequency cancellations spaced harmonically. As the delay time is modulated, these frequency cancellations sweep dramatically up and down the spectrum, creating a psychedelic swooshing sound like the sound of an aeroplane taking off and landing repeatedly.

Chorus, on the other hand, uses delay times of approximately 5 ms to 25 ms to create a thickening of the sound, giving a false impression of many different instruments (a chorus). The modulation is typically less dramatic than flange, although it is worth noting that many flange effects can create chorus, and many chorus effects can create flange.

2.2.3.2 Phasor

Phasor or phase shift effects work similarly to flange in that the signal is duplicated; but instead of delaying the duplicate signal, the phase of the carrier is shifted and modulated. Phasor is typically a more subtle effect than flange, but is still common in psychedelic music.

2.2.3.3 Tremolo

As described in Section 2.2.3 and shown in Figure 2.16, tremolo is the effect of modulating the amplitude of a carrier with an LFO. This low frequency modulation of volume causes a trembling or wobbling texture to the sound – which, like spring reverb, is very popular with electric guitars, especially in the surf rock genre. It is worth noting that due to initial mislabelling, the 'tremolo' arm on a guitar does not create tremolo, it creates vibrato. Conversely, many guitar amps – including the famous Fender twin – have a vibrato channel, which, confusingly, is actually tremolo.

2.2.3.4 Vibrato

Vibrato, as introduced in the previous section, is the effect of modulating the *pitch* of a carrier with an LFO. Vibrato is also a very common performance articulation in pitched instruments that gives a note a dramatic and expressive sound.

2.2.3.5 Auto wah

As the name suggests, the auto wah effect is an automated version of a wah pedal effect. The main element of a wah pedal is a resonant low or bandpass filter, which has its cutoff frequency swept using a rocking pedal. The auto wah uses either an LFO, an envelope, or a combination of the two to modulate the filter's cutoff. An envelope in the context of an audio effect like an auto wah typically uses transient detection to identify the attack sustain and release of an audio signal. This signal rise and fall then modulates the filter's cutoff. Auto wah creates a rhythmic funk sound that was particularly popular on electric guitar, bass guitar, and electric pianos in the 1970s.

2.2.3.6 Vocal effects

While any effect can be applied to vocals, specific types of effects can be used to subtly correct and colour a vocal take or to add a more synthetic sound – for example, the characteristic 'robot voice' type effects used widely in electronic music, specifically in the vocals of the song 'Harder, Better, Faster, Stronger' by Daft Punk.

2.2.3.6.1 Autotune/pitch correction

Autotune is an example of a sound manipulation technique initially intended to correct the intonation or tuning of a vocal take in a corrective rather than creative manner. The incoming vocal is fed through a pitch detection algorithm and is then pitch shifted to match the closest note on a desired scale (pitch shifting is discussed in Section 2.3.1.5.3). If the vocal is autotuned aggressively, the transitions between notes become artificial and stepped sounding, leading to the effect made famous by the Cher song 'Believe' and used widely as a creative effect, particularly in modern R&B music.

2.2.3.6.2 Vocoder

The name 'vocoder' is a combination of the words 'vocal' and 'encoder'. Like many other sound manipulation techniques, it is a musical application of telecommunications technology. The vocoder was designed to reduce the bandwidth required for transmitting a human voice signal. It did this by measuring the changing frequency spectrum using a set number of frequency bands and then transmitting only the amplitude envelopes of each frequency band. This information was then used to modulate a complex carrier signal (e.g., white noise), allowing an intelligible recreation of speech at the receiver, albeit with a robotic voice.

As an effect in music production, the encoded information from a vocal is used to modulate a musical carrier – for example, a synthesiser, resulting in a 'singing synthesiser' sound. A good example of a vocoder in music is the vocal in 'Mr Blue Sky' by the Electric Light Orchestra.

A similar process of producing vocal sounds that also comes from telecommunications technology is linear predictive coding (LPC). LPC, like the vocoder, is a means of dramatically reducing the bandwidth needed to transmit speech, which is, in some ways, the digital counterpart to the analogue vocoder. LPC uses a simplified model of human speech that assumes speech sound is made up of a buzz created by the vocal cords, which is then manipulated by the size and shape of the vocal tract, creating formants (resonant frequencies). LPC estimates the formants of the vocal tract and removes these sounds to estimate the level and frequency of the buzz. The buzz is then recreated using an oscillator and sent through a formant filter to create an artificial but intelligible speech signal. The Speak & Spell toy of the 1980s is a good example of a popular application of LPC.

Both the vocoder and LPC are examples of resynthesis: the process of recreating a sound from key descriptive elements. Resynthesis is discussed in the context of synthesis in Section 3.6a.8.

2.2.3.6.3 Talkbox

A talkbox is an amplifier and driver inside a box with a hole connected to a tube. The box, amplifier, and driver cause significant distortion. The sound travels up the tube and into the performer's mouth, where the mouth imparts the vocal formants projecting the signal into a vocal microphone. Typically, the instrument signal connected to the talkbox is a guitar or synthesiser, which can be considered the carrier. The modulator is the shape of the mouth, which acts like a physical vocoder. Notable examples of the talkbox include 'Show Me the Way' by Peter Frampton and 'Livin' on a Prayer' by Bon Jovi.

Talkbox effects typically emulate the distortion of the box and the application of the formants by extracting the envelopes of the vocal formants and applying them to the carrier signal. The talkbox effect can be similar to a distorted vocoder effect.

An example of a plugin that uses these effects together is VocalSynth 2 by iZotope. A screenshot of the VocalSynth 2 plugin is shown in Figure 2.17. The Polyvox module is a harmoniser (discussed in Section 2.2.6), and the Compuvox module is an application of LPC. (The Biovox module estimates and manipulates

FIGURE 2.17
iZotope VocalSynth 2 plugin (cropped to remove effects)

vocal tract shapes to change formants and tone and the Pitch controls adjusts pitch detection and correction settings.)

2.2.4 Distortion

In audio, distortion is a signal alteration, typically occurring when a device overdrives. Overdrive occurs when a device is driven to the point that the signal can no longer go louder in a linear fashion (or at all), so the signal is altered.

Specifically, this alteration is likely to be a combination of compression and the addition of overtones. Sometimes this alteration is not desired – for example, in public address (PA) systems for live sound where distortion is likely to reduce the clarity, or in digital audio systems where a recorded signal is clipped, resulting in harsh square waves. However, in other cases, especially in guitar amplifiers, the sound of a device overdriving can be a highly desirable effect – so much so that there are a significant number of different types of distortion, with many creative uses far beyond the electric guitar.

The different terms for a device overdriving range from the colloquial 'going into the red' to the more specific 'clipping', which happens when a system reaches its absolute maximum, and the top part of the waveform is flattened. As distortion is typically created by non-linear amplification, a degree of dynamic range compression is also inherent in the effect. A general summary of this overlap between distortion and compression devices is that most distortion effects result in dynamic range compression, but relatively few compressors result in noticeable distortion (with the notable exception of the 1176 compressor, discussed previously).

2.2.4.1 From overdrive to distortion

In audio effects, the term 'overdrive' comes from early electric guitar blues amplifiers. When the vacuum tubes and loudspeaker were driven to overdrive, they produced new frequencies, often harmonic, adding to the tone to give a warm and dirty characteristic. A trait of vacuum tube amplifiers is that they do not immediately clip (as transistors tend to), but gradually reduce amplification or give soft clipping. This soft clipping gives a warm, growling tone, typically accentuating the second and fourth harmonics, which sound more musical (as they are octaves of the fundamental note) than odd-order harmonics (which are different musical notes).

Although 'distortion' is also used as a general descriptor and can be used interchangeably with 'overdrive', in the specific context of audio effects (especially electric guitar effects), 'distortion' refers to a more pronounced and aggressive effect than overdrive. Distortion gives the characteristic heavy rock and metal guitar tones for electric guitars. Vacuum tubes are still very popular for electric guitarists as the complex way the vacuum tubes react to the incoming signal is very hard to model authentically. However, some digital amplifiers, amp modelers, and plugins are edging ever closer to the tones created by vacuum tube amplifiers.

2.2.4.2 Guitar amplifier simulators

Due to the convenience of direct input recording and the flexibility of setup and preset switching, there are many guitar amplifier simulator plugins. As well as plugins (and the actual amplifiers themselves), there is a wide range of amplifier modelling pedals that do away with the need for large amplifiers and cabinets.

It is also worth noting that due to the very mid-heavy frequency response of electric guitar pickups, the tone controls of guitar amplifiers and effects are typically biased to accentuate bass and treble frequencies (the typical loudness, or 'smile' EQ). Because of this, guitar amplifiers and effects can excite signals other than electric guitars.

2.2.4.3 Saturation, waveshaping and drive

Beyond guitar effects, distortion has found many uses in the broader area of music production and goes by several other terms depending on its use and application – for example, 'saturation'.

Saturation is typically used to convey a more subtle addition of harmonic content, such as the effect added to a high-level signal stored on magnetic tape or the tonal colour of preamps from outboard equipment or analogue mixing desks (most notably Neve and SSL).

Waveshaping is a prescriptive method of specifying exactly how a wave will be altered depending on its negative or positive amplitude. The user interface will typically use a bipolar transfer curve: a graph showing an input in the vertical x-axis and output in the y-axis, showing both negative and positive areas of the wave. A Waveshaper can be symmetrical – that is, it applies the same variation to the positive and negative elements of a wave, or it can be asymmetrical so that the alterations to the positive and negative parts of a wave are different. The bipolar transfer function of the Ableton Live effect Saturator is shown in Figure 2.18, showing symmetrical waveshaping. (Figure 2.6 for the Ableton Live compressor and Figures 2.10 and 2.12 for the Pro Tools Expander/Gate are single-pole transfer functions because the signal level is positive only.)

FIGURE 2.18
Ableton Live Saturator plugin showing bipolar transfer function

Several other devices can create desirable distortion, especially filters – often by using the drive, gain, or feedback controls.

2.2.5 Reamping

Reamping is the direct recording of an instrument, such as an electric guitar, bass, or keyboard, which is then played back through an amplifier and re-recorded with one or more microphones to replace or mix in with the direct sound. Reamping allows a broader range of experimentation than the traditional method of playing into one amplifier and recording that 'live'. It is worth noting that the output of an audio interface is likely to have a much lower impedance (or 'Z') than an electric guitar. Consequently, when plugged into an electric guitar amplifier (or other device designed for high Z input), the mismatched impedance may change the tonal characteristics (i.e., it may lack high frequencies). If needed, a solution to this is a reamp box (or a passive DI in reverse), which gives a high Z output.

The software-only version of reamping uses an amplifier simulator as described in Section 2.2.4.2, which significantly simplifies the reamping but is likely to lack the unique character of a physical amplifier and microphone combination.

Another variation of reamping is to play back any signal (not just one recorded directly) and re-record the amplified sound. While not an effect in itself, the result of reamping is a wet and dry track to mix and can be an effective method of bringing additional character – especially when working in the box, which can tend towards a clean and clinical sound.

2.2.6 Harmonisers

A harmoniser is a device that pitch shifts in real time and allows the pitch-shifted signal to be mixed in with the original, giving an artificial harmony. One of the earliest and most characterful harmonisers is the Eventide H910 Harmonizer®, released in 1974.

The plugin version of the Eventide H910 Harmonizer is shown in Figure 2.19. A harmoniser is the real-time equivalent of the offline technique of pitch scaling, discussed further in Section 2.3.1.5.2. Because harmonisers work in real time,

FIGURE 2.19
Eventide H910 Harmonizer® plugin

they are more likely to cause audio artefacts and colour the sound, whereas off-line pitch scaling is likely to give more transparent results.

2.3 EDITING

Editing has been, and continues to be, a technique vital to the development of music production. Early editing examples include rearranging musical parts on magnetic tape with a razorblade. Editing affords dramatic improvements to musical recordings; however, editing on tape can be particularly unforgiving of mistakes. As digital technology has become ever more powerful and accessible, the ease with which even the tiniest audio element can be transformed has fundamentally changed how we experience music and sound. While editing is core to this development, so are the sampling, synthesis and Music Instrument Digital Interface (MIDI) technologies that have evolved from and alongside audio editing. (Sampling and synthesis are discussed in Chapter 3.)

MIDI (introduced in Section 1.1) is data often used to trigger and control electronic musical instruments, much like a musical score instructs a musician to play notes of a certain length. MIDI editing differs significantly from audio editing in that MIDI allows far more control over the notes' performance and the instrument's sound. MIDI editing is discussed in Section 2.3.2.

Also included in this section are the offline processes available in the recording studio, such as reverse, time stretch and polarity invert.

2.3.1 Audio editing

Editing audio in the recording studio can involve removing unwanted audio, rearranging audio parts, or re-drawing the waveform. Editing can also be rendering – that is, the offline processing of an audio file. Sections 2.1 and 2.2 discuss processors and effects as real-time audio manipulation – that is, the audio is manipulated as it is played back, allowing a considerable degree of flexibility in changing parameters, bypassing and reordering. However, some audio manipulation techniques are impossible to operate in real time and rendering to the audio file is more desirable. Offline editing is discussed in Section 2.3.

There are two levels of digital audio editing available in audio software: clip (or arrangement) editing and sample (or waveform) editing.

Clip editing: A clip is a virtual container for musical material, either MIDI or audio. Clip editing is a form of non-destructive editing, which means that if an audio clip is edited, the audio file itself is not changed. The clip refers to all or part of an audio file. The clip that is viewed and edited represents instructions to play and possibly modify (e.g., fade-ins and fade-outs). Because the audio file itself is unchanged, it will be processed every time it is played back to apply the clip parameters.

Sample editing: Sample editing relates to the direct editing of the sample – that is, the audio file itself. Sample editing can be either destructive or non-destructive. If the sample editing is destructive, the changes being made are happening directly to the audio file and, when saved, are likely to overwrite the original file (if the name is not changed). If the sample editing is non-destructive, then every edit will cause a copy of that audio to be saved, ensuring that there is a backup if needed. Accordingly, destructive sample editing is the fastest and most computationally efficient method of sample editing, while non-destructive sample editing is far safer.

Clip editing is the method of audio editing used by default by most modern DAWs due to its flexibility and safety of operation, and the relative abundance of processing power and computer memory available in current systems. Most DAWs have a sample editing capability, but this area is typically separate from the main arrangement view and is likely to be used for the speed of workflow and file management reasons necessary, for example, in long-form, radio or podcast creation. Ableton Live is notable as a DAW that does not have a sample editor (as of version 11).

2.3.1.1 Non-zero-crossing artefacts

An audio waveform is a graphical representation of the displacement of air molecules above and below a zero line. The waveform is equivalent to the position of a loudspeaker cone if the loudspeaker is placed on its back facing upwards (as illustrated in Section 1.1.1.1). When creating sound at its loudest, a loudspeaker travels upwards to its maximum point, back through the zero line, and then to its minimum point. The waveform continues back and forth with varying degrees of symmetry to create sound. The loudspeaker rests at the zero line when the sound fades out to silence.

When audio is edited, the waveform is split. If this split occurs at a point that is not on the zero-crossing, the speaker will attempt to travel from the edit point to the zero line immediately, which is likely to create a high-frequency artefact in the form of a pop or click. An example of a waveform edited on a non-zero crossing is shown at the start of the clip in Figure 2.20. This edit creates a hard electronic 'tick' sound. An artefact can also occur even if the edit is on a zero-crossing but at a high energy point, which causes the loudspeaker cone to attempt an unnaturally abrupt stop.

To avoid or remove these artefacts, many DAWs offer an option to edit only on zero-crossings or have a setting to create tiny fade-outs on clip edges

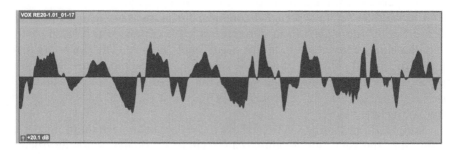

FIGURE 2.20

Audio clip waveform with an edit on a non-zero crossing in Pro Tools

automatically. (In Pro Tools: Preferences…>Operation tab> Misc section "Clip Auto Fade In/Out Length" in msec; in Ableton Live: Preferences>Record Warp Launch tab>Warp/Fades section "Create Fades on Clip Edges" on or off.) Either way, it is good practice to use caution when editing and manually create fades to avoid these artefacts. It is also worth noting that in some cases, a harsh edit may be part of the aesthetic – the most obvious example being glitch music, a subgenre of popular electronic music (and what may be categorised as intelligent dance music).

2.3.1.2 Silencing and silencing tools

Removing unwanted audio, otherwise called silencing, can be done by manually editing out the unwanted sections (while ensuring that there are no non-zero-crossing artefacts, as discussed in Section 2.3.1.1). The manual approach to this task can be better when there are only a few sections to silence or when each section benefits from a more flexible approach. Otherwise, this can be a particularly laborious task and well suited to the meticulous nature of a computer in the form of silencing tools, such as Strip Silence in Pro Tools.

Silencing tools are beneficial for quickly and conveniently removing many audio sections using the same parameters. Figure 2.21 shows a screenshot of the Strip Silence window in Pro Tools with a selected clip below. The selected

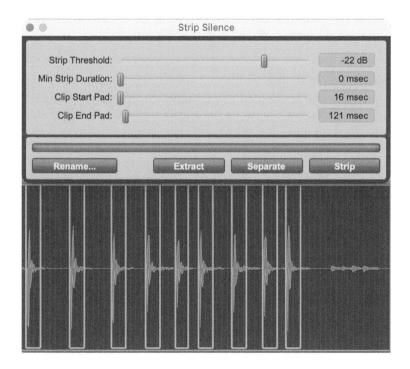

FIGURE 2.21
Strip Silence window and selected audio clip in Pro Tools

clip shows grey boxes overlayed on the waveform previewing the clips that will remain when the strip option is used.

Silencing tools can also be helpful as a straightforward method for batch adjusting the attack and release transient characteristics of percussive sounds – for example, close-miced drum signals.

The two significant benefits of silencing tools over gates (see Section 2.1.4.1) are that different sections can be given different parameters; and the results of each silencing operation can be seen, previewed, and altered. Silencing tools afford a far more dramatic and clinical option than gates and are especially well suited to signals from close-miced drums. Because silencing tools can so efficiently remove any potentially unwanted low-level signal elements, this allows for far more compression, as there is less likely to be the muddiness and spill that would be boosted otherwise.

2.3.1.3 Comping

Comping is recording several takes of a music part and selecting the best sections from each take to compile a composite track (the term 'comping' comes from either 'compile' or 'composite'). Most DAWs that are designed with a focus on recording audio have comping tools built into the audio tracks to be used as needed (e.g., playlists in Pro Tools, take folders in Logic and take lanes in Ableton Live). Comping tools, combined with the time stretch and pitch scaling tools discussed in Section 2.3.1.5.2, allow musical parts to reach an obsessive level of clinical perfection; but this also means that using these tools to this degree may remove much of the expressive feel from a performance.

2.3.1.4 Phase alignment

When a sound source is recorded using multiple microphones that are different distances from that source (e.g., a kick drum microphone inside the drum and another microphone outside), the signals will have a tiny delay which, when combined, can cause undesirable phase issues.

Phase is a delay or offset of a signal, measurable as a degree of the waveform (from 0° to 360°). When two signals are combined with a phase difference, this may reduce many of the frequencies and boost other frequencies, potentially resulting in a thin and unpleasant sound. Waveforms can be aligned to match the phase of some frequencies, but it is unlikely that all signal frequencies can be phase aligned. Phase aligning can be done by zooming in on one of the waveforms and nudging it to line up with the other so that the most critical frequency comport can be matched. As with most audio manipulation, phase adjustments should be judged with a critical ear, as it might be that the phase difference provides a worthwhile characteristic. Some plugins can automatically align phase – for example, Auto-Align by Sound Radix; however, they are less common than silencing tools.

It is worth noting that a signal that is the same as another but exactly 180° out of phase is likely to disappear almost entirely. This is described in the context of balanced cables in Section 1.2.1.2.1.

2.3.1.5 Rendering

Rendering is when any process is committed to an audio file, either rewriting an original or creating a new audio file. As discussed in Section 2.3.1, the previous sections in this chapter discuss audio manipulation techniques in terms of real-time processors and effects. There are, however, some techniques that cannot work in real time on live audio and situations where rendering may be more desirable in terms of workflow, system efficiency, and allowing further editing of rendered audio.

Rendered audio could be used as an effect on an additional track; however, this is not likely to be effective, as phase issues will be expected and somewhat contrary to the point. Typically, rendered audio replaces the original, so they are discussed here as processors. Real-time processors and effects (discussed in Sections 2.1 and 2.2, respectively) can often be rendered, but to reiterate the other side of the point made in Section 2.3.1, this removes any chance to change parameters, to bypass, or to re-order.

The audio manipulation techniques that cannot operate in real time on live audio are reverse, time stretch, pitch scaling, and normalisation. These techniques cannot work in real time because they would need to look into the future to operate in real time with live signals. Reverse is discussed in Section 2.3.1.5.1, while time stretch and pitch scaling are discussed in Section 2.3.1.5.2. While pitch shifting can be done in real time, devices that do this are typically called harmonisers and are discussed in Section 2.2.6. The requirement for harmonisers to manipulate audio nearly instantaneously so that they can be used in real time means significantly more chance of colouring the audio or sounding artificial compared to offline pitch shifting. From this point on, the term 'pitch shifting' is used to refer to offline pitch shifting, which is discussed in Section 2.3.1.5.3. Normalisation is discussed in Section 2.3.1.5.4. Polarity inversion (represented by the symbol ø) is included here as a utility, which is often used as an offline process for simplicity and convenience (although plugins can invert polarity too), and is discussed in Section 2.3.1.5.5.

2.3.1.5.1 Reverse

Reversing audio is the straightforward task of taking an audio file and playing it backwards so that the end is played at the beginning and vice versa. Reversing audio is, of course, impossible to do in real time with a live source (however, it is possible to buffer a short section of audio and reverse that live). As well as the psychedelic 'shoop' sound of reversed audio, an established experimental approach is to reverse the audio, apply reverb, and then reverse again so that the original signal now plays as before. The result is that the reverb now precedes the sound, giving a strange, time-defying reverb effect.

2.3.1.5.2 Time stretch and pitch scaling

Time stretch is the process of changing the duration of recorded audio without any change to its pitch, which results in the speeding up or slowing down of a recorded audio part. The typical corrective use for time stretch is manipulating a musical phrase to synchronise with other music. Along with the increased availability of greater computer processing power, time stretch manipulation has become embedded into many DAWs – for example, the warp functions of Ableton Live and Elastic Time in Pro Tools. These functions allow highly powerful and sophisticated correction of audio parts, which can completely transform the timing of an audio recording in a way that can be difficult to detect (particularly in the mix).

When used to the extreme and modulated or automated, time stretch can create digital artefacts that can be used as a creative effect in more experimental forms of popular electronic music, giving a digital and sometimes glitchy sound.

Pitch scaling is the opposite process of time stretching – that is, changing the pitch without any change to duration. The most common application of pitch scaling is autotune, where a musical part (most commonly vocals) has its pitch corrected to a musical scale. As with time-stretching tools, pitch scaling tools have reached an astonishing level of sophistication. It is common to find both pitch scaling and time stretching available in programs such as Melodyne by Celemony or Autotune by Antares, which can be used within a DAW as a plugin or a standalone application.

While it is a less common technique, audio can be captured live and processed when looped – for example, the live looping scenario described in Section 1.2.3.1. Time stretching techniques are central to live systems that allow tempo variation. Live techniques to manipulate audio and MIDI are explored in Chapter 5. These are critical parts of systems that turn the non-live elements discussed in Section 1.2.3 into pseudo-live elements.

2.3.1.5.3 Pitch shifting

'Pitch shifting' is a term used to describe either the real-time live process of changing the pitch of a sound or the offline changing of pitch (making it interchangeable with pitch scaling in that situation). The distinction between real time and offline is often indicated by context: if the device is an insert effect, then it is operating in real time; while if the device is a process applied to a clip, it is offline.

Because of the speed required to achieve real-time pitch shifting, the algorithms used are likely to sound more artificial and cause artefacts than the offline counterparts. Accordingly, it is unlikely that a real-time pitch shifter will produce more desirable results than offline pitch shifting (or pitch scaling) when working with clips.

2.3.1.5.4 Vari-speed pitch control

A variable-speed pitch control, or vari-speed control, is a way to change the pitch of a sound by slowing down or speeding up recorded audio; so, unlike pitch scaling, the duration *is* changed. Common examples of pitch controls include the pitch control slider on the side of a turntable that speeds up or slows down the speed of the vinyl, or a speed adjustment on a reel-to-reel tape recorder. It is also worth noting that speeding up or slowing down the playback of a sample is the typical method of pitch adjustment used by samplers when a sample is spread over keys that move away from its root position.

2.3.1.5.5 Normalisation

Normalisation is the process of scanning an audio file to identify its highest peak and then raising the gain of the whole file so that the highest peak is set to a ceiling value in dB. The highest ceiling value is 0 dBFS; however, 0 dBFS is the absolute resolution limit, so to avoid any clipping, the ceiling value should be below 0 dBFS – for example, -0.3 dBFS. Normalisation is done to make an audio file as loud as possible without changing its dynamic range and make the most of a system's resolution. It is most effective when applied as the final stage in producing audio. It is also common practice to normalise as the final stage in making sample packs or for a backing track. However, it is worth noting that the downside to having such strong signals is that, if combined with other sounds, the gains of each strong signal will need to be reduced significantly to avoid clipping.

A standard misuse of normalisation is to make samples louder within a DAW project, mainly due to the ease of use and for immediately loud results. However, using a gain device is better practice, as it gives more control.

2.3.1.5.6 Polarity inversion

As mentioned in Section 1.2.1.2.1, polarity inversion swaps the positive waveform with the negative waveform. Polarity inversion, or the 'ø' switch, is often used on the signal from a secondary microphone on a single source pointing in the opposite direction – for example, a microphone underneath a snare pointing upwards.

Although polarity inversion is easy to achieve, as it is included in many gain or utility plugins as an insert, it is typically more helpful to free up an insert by processing the track offline.

2.3.1.5.7 Spectral editing

Spectral editing is a highly in-depth type of audio editing where audio is typically shown and edited using a detailed version of a spectrogram. A spectrogram shows the frequency spectrum as different colours in the vertical y-axis and time in the horizontal x-axis. Good examples of software for spectral editing are SpectraLayers by Steinberg and RX by iZotope.

2.3.2 MIDI editing

As introduced in Section 1.1, MIDI is control data, often used to control electronic musical instruments, much like a musical score instructs musicians to play notes with a certain duration. At the most basic level for musical note generation, the MIDI message starts with a note on message consisting of pitch and velocity information and ends with a note off message (or a note on message with the same pitch and a velocity of 0). The pitch is the note played (i.e., MIDI note 60 – middle C, 262Hz), and the velocity is how loudly to play the note (i.e., MIDI velocity 100 – quite loudly). A more detailed discussion of MIDI messages, including MPE and MIDI 2.0, is set out in Section 4.2.

Despite the advances in time stretching and pitch scaling discussed in Section 2.3.1.5.2, editing MIDI is far more flexible and convenient than editing audio, as MIDI is control data and allows extensive control over the notes that are either performed or programmed. Editing MIDI information is a step away from actual audio/sound creation so that any available parameter can be easily changed.

Although some sound characteristics are changeable with audio editing, this is more of a correction process than simply changing the sound parameters. For example, to change the pitch of a MIDI note up a perfect fifth, that note can easily be moved up the piano roll editor by seven semitones, and the resulting note will be sounded with just as much tonal purity as the original note. The pitch of an audio note can be moved up a perfect fifth using a pitch shift plugin or the vari-speed pitch control of a sampler. However, either process will make the tone sound artificial, whereas the MIDI note will sound just as authentic as the previous pitch.

2.4 STEREO IMAGE

Before the late 1950s, all commercially released music was in the mono format: one music channel. Two-track, stereophonic or stereo releases followed, giving two different channels, each to be sent to a separate loudspeaker providing a 'stereo image'. Despite the efforts of surround consumer formats (quad in the 1970s, 5.1 in the 2000s and spatial audio in the 2020s), stereo has remained the standard format for commercially released music for over half a century. Accordingly, the stereo image of a mix is a vital aspect of mixing music.

2.4.1 Panning

Any signal sent to both loudspeakers sounds like it is coming from somewhere in the middle. This phenomenon is called the phantom centre. Early mixing desks worked in LCR, which meant the signal could only be sent to the left, right, or both channels (the centre). Following LCR mixing, the pan pot ('pot' is short for potentiometer, a variable resistor) came into use. In its centre position, the pan pot sends the signal equally to both the left and right channels. As the pan pot is turned anti-clockwise, the signal's volume in the right channel is reduced, making the sound appear to be coming from the left. Similarly, as the pan pot is turned clockwise, the signal's volume in the left channel is reduced, making the sound appear to be coming from the right.

Panning is the standard method for placing mono signals in the stereo field. For signals recorded using spaced microphone techniques, the stereo image is based on the signal's timing and the amplitude of the signal captured by each microphone. The basic standard conventions for instruments in the stereo image are worth knowing so that these can be adhered to or broken as desired. A straightforward summary of basic stereo image conventions for most popular music genres is main vocals, bass and kick drum in the centre, and accompanying instruments with significant mid and high frequencies balanced across the stereo field.

2.4.2 Haas panning

In contrast to using the relative volume of the signal in left and right channels, and similar to how spaced microphone techniques work, an alternative method for moving signals in the stereo field is to use Haas panning. Haas panning works by creating a copy of the audio, panning both signals hard left and right and then delaying one of the signals. If the delay is between 0.01 and 0.07 ms, this results in the perceived movement of the sound across the stereo field, with 0.07 ms being the point at which it sounds hard-panned. Any more than 0.07 ms is likely to thicken rather than an apparent movement across the stereo field. While Haas panning is rarely used to control the stereo position of respective sources, it maintains an equal amount of signal power in each channel.

2.4.3 Stereo in a live environment

In a live environment, when using a PA system, sometimes the PA is run in mono (or stereo bridge, when both channels are combined) mode. A mono PA can be beneficial – partly because the mix is likely to sound more powerful with both channels playing the same signal, but also because any dramatic panning results in a large proportion of the audience missing out on instruments panned away from them. Any stereo image that is used in a live environment is likely to be significantly more subtle than the stereo image created in the recording studio so that the audience members to the side do not miss out on any significant musical elements.

2.5 PRODUCTION OF NON-LIVE MUSICAL ELEMENTS FOR LIVE PERFORMANCE EVENTS

The backing track is the most common form of non-live musical element for live popular music performance events. Other less common forms include sample banks, audio clips, and MIDI clips.

This section discusses the production of these standard non-live elements in the established context of fixed musical aspects of a live performance event. These ideas are then expanded and applied to create and manage pseudo-live elements in Chapter 5.

2.5.1 Backing track production

A backing track is a fixed audio file played so musicians can play along. As discussed in Section 1.2.3.1, a backing track offers the potential of a nearly infinite expansion of sounds in live performance. The downside is that the musicians must synchronise to the backing track, typically using its accompanying click track.

The traditional method of playing back a backing track is to press play on a good-quality two-channel media player. The backing track consists of a mono front-of-house (FOH) mix sent on one of the channels, and a cue mix (typically a prominent click and the backing track) on the other channel, to be sent to stage monitors. This two-channel media player arrangement is likely the most reliable and cost-effective method, making it a valuable backup to a more complicated setup.

If multi-channel (i.e., more than two channel) backing tracks are required, then a computer running a DAW such as Ableton Live (along with a multi-channel audio interface) or a dedicated multi-channel media player may be used. The channels making up a multi-channel backing track can be anywhere from a limited set of stems (sub-mixes – for example, drum mix, strings, and backing vocals) to an array of discrete channels that may look similar to a recording studio setup. A limited stem option is likely to be easier to set up and more reliable, so it would be an effective option for a fixed backing track where a more flexible mix is needed for FOH and monitors. If the backing track is to be manipulated in any way more than just balance, many DAWs afford the flexibility and power to do this. Techniques for augmenting backing tracks are explored in Chapter 5, among other approaches to augmenting live performance.

While the range of audio manipulation techniques is the same for both backing tracks and a track for release in use, backing tracks may need to integrate with the audio from the live performance (if this is the desired aesthetic). However, this is unlikely if they are overhyped by the techniques discussed earlier in this chapter. Backing tracks are most likely to integrate when there is a balance between making the backing track sound polished, consistent, and exciting and maintaining a large enough dynamic range of a backing track to sound like an integral part of the performance.

2.5.1.1 Dynamic range in live and non-live elements

As mentioned in Section 2.1.2, recorded (i.e., non-live) music has been made to sound ever louder by dynamic range reduction, leading to a situation often referred to as 'hyper-compression'. However, the audio quality of hyper-compression afforded by a recording studio is almost impossible to achieve with live elements in a live performance situation. As discussed in Section 2.2, the make-up gain applied after compression causes low-level signals to get significantly louder, but in live sound, this makes howling feedback far more likely for live signals going through a PA (see Section 1.2.2.1 for a description of howling feedback).

Consequently, the relationship between the dynamic range of the live elements and the dynamic range of the non-live elements – typically backing tracks – is worth careful consideration. These dynamic ranges may not necessarily need to be matched, but their difference significantly determines the performance aesthetic.

Non-live elements are often created from musical parts captured in a recording studio or electronic music environment for a musical release. In this situation, the parts are likely to have a significantly compressed dynamic range. Depending on the desired sound, it may be necessary to reduce the compression by removing some compressors or raising their thresholds and lowering their ratio. If the signal is already bounced (i.e., rendered to an audio file), then the dynamic range of the audio can be increased by careful use of an expander (see Section 2.1.4.2).

One method to test the difference between the dynamic range of the live signals and the dynamic range of non-live elements is to set up a performance and record the live parts with only the processors and effects used live. These live test signal(s) can be auditioned alongside the non-live elements, which can then be adjusted accordingly and tested in a live situation (which can be re-recorded if further refinement is required).

2.5.1.2 Corrective editing

The opportunity to record multiple takes, comp these takes (discussed in Section 2.3.1.4), and manipulate the pitch and phrasing means that the parts have clinical perfection that is impossible to recreate consistently in live performance (similar to how dynamic range manipulation in a recording studio affords a potentially vast and explosive sound). This clinical perfection may, of course, be part of the performance aesthetic. However, it is worth considering how these perfect takes combine with the live performance. For a typical backing track, if accompaniment is more varied and textural (and has a wider dynamic range), it is likely to blend more smoothly into a live performance.

2.5.1.3 Use of effects in the production of non-live elements

Live effects, as with live processors, tend to be fewer and less complicated than the effects available in the recording studio. Most signals from microphones in an on-stage environment are likely to be relatively dry (i.e., have little ambience and no effects added), with the notable exception of creative effects integral to the character of instrument sounds, such as spring reverb on electric guitar. Accordingly, if the non-live elements are kept relatively dry, they can be sent to the same effects as the live signals to blend more effectively.

2.5.1.4 Tempo track manipulation for backing tracks

Most backing tracks are made to a fixed tempo, and while this gives a rhythmic consistency that may suit some types of music, it also dramatically limits the degree of expressiveness and movement.

More rhythmic interest and movement can be added by creating a tempo track that speeds up and slows down a DAW where appropriate. While this is, to a certain extent, 'faking' a piece of music that has been played with a natural intended push and pull, the music will be less static.

One method for creating a tempo map is to record an exemplary live performance used as a tempo template that can be refined. Most DAWs have tools that can aid in mapping performance timing onto a tempo track (e.g., Beat Detective in Pro Tools and Time Warp in Cubase).

These tempo changes are still fixed, though, and if the same tempo maps are used repeatedly, they may be just as tiresome to perform as a fixed tempo.

Alternate backing tracks can also be created with varying tempo changes (with different dynamics and phrasing if required). However, this does require a significant degree of trust in the person choosing the backing version and a keen ability to judge the performance.

2.5.2 Samples as non-live elements

Similarly to the use of backing tracks, samples dramatically open the sonic opportunities of performance events. A detailed exploration of sampling concepts and techniques is covered in Chapter 3. This section introduces sample production considerations in the context of the recording studio techniques discussed earlier in this chapter.

In live use, there are two extremes of sample production: the authentic recreation of an instrument, and the more experimental samples that use sonic manipulation tools, and possibly the process of capturing unique sounds. There are, of course, a wealth of sonic possibilities between these two extremes.

Authentic recreation samples are often large sample libraries created and sold as electronic instruments. These sample-based instruments are carefully edited and velocity-layered, sometimes using data compression to manage disk space. They are designed to sound as close to a professionally recorded instrument without the difficulty of playing and recording. (However, along with this convenience is the lack of control and uniqueness that comes with a live recording.)

Typically, sample instruments (as with synthesisers) are triggered/controlled in a live performance environment by MIDI controllers. Percussive samples are likely to be triggered by drum pads, which are often standalone devices sat next to the snare of a drum kit, or pad-based controllers to be played using fingers. Pitched samples are typically triggered using a piano keyboard. MIDI controllers are discussed in further detail in Section 4.3.2.

The more experimental samples can be a standard sounding sample like a kick drum, or a vocal part, which is manipulated to sound like something dramatically different. Sampling techniques are discussed further in Chapter 3.

One of the sonic strengths of samples is their creative flexibility. They can be edited, processed, layered, have effects added, and then be resampled (rendered for more manipulation), and this process can be repeated over and over. However, similarly to the produced backing tracks discussed previously, the sound of this manipulation can be hard to fit into a mix, as it is likely to swamp many acoustic and electric sources.

2.5.3 MIDI for non-live elements

As discussed in Section 2.3.2, MIDI messages are data that is used to control music or sound-creating devices and, as such, are highly flexible and editable. The basic details of MIDI 1.0 messages are discussed in Section 4.2.

The considerations for producing MIDI for non-live elements (i.e., the use of playback in live performance) are very similar to those considered when producing an audio backing track if it is played back in a fixed fashion. The main benefit of using MIDI tracks over audio tracks is the significantly greater degree of flexibility that MIDI offers. However, in the development of technology – particularly in the early stages – flexibility comes at the cost of reliability. Because of this, the standard for live accompaniment in popular music has become entrenched in the single audio file backing track.

MIDI as a live interface standard (rather than a storage standard – that is, MIDI information stored on a MIDI/instrument track) is particularly well suited to trigger and manipulate sounds in a live environment. While MIDI tracks can be used to make up a fixed backing track, there is little point if the backing track is fixed. More sophisticated backing systems are discussed in Chapter 5, where MIDI plays a significant part.

Electronic sound creation techniques: Synthesis and sampling

A s discussed in the previous chapter, there are two main paradigms for music production: the recording studio and the electronic music environment. These paradigms are not at all separate; they coexist, overlap, and cross-fertilise, affording an ever-evolving array of creative practices. The previous chapter discusses the sound manipulation techniques of the recording studio, which can be summarised as 'recording' music. This chapter discusses the methods and tools of electronic sound creation, which can be summarised as 'building' music. The specific techniques of electronic sound creation are the techniques of synthesis and sampling.

The word 'synthesis' is used widely in spoken and written English, covering many different contexts, from chemical synthesis to educational theory. In audio, the term 'synthesis' means 'sound synthesis' – that is, electronically generating and manipulating audio signals to create musical sounds.

Similarly, the term 'sampling' also has a broad range of meanings. However, in the context of audio, the term 'sampling' means the recording, playback, and possible manipulation of audio.

The general difference between synthesisers and samplers is that synthesisers typically generate the initial tone and almost always use sound manipulation techniques to transform this initial tone into a rich and complex sound. On the other hand, samplers use recorded sound – or samples – as a method of sound creation. While they are often capable of the same sound manipulation as synthesisers, their use is not inherent in the sound creation process.

As will become apparent when discussing specific techniques later in this chapter, there are instances of significant overlap between concepts of synthesis and sampling. As a general rule, because they often generate sound from scratch, synthesisers are often well suited to creating new and different sounds rather than authentic recreations of acoustic instruments. On the other hand, as they are based on recordings, samplers are often well suited to authentic recreations

 DOI: 10.4324/9781003370406-4

of acoustic instruments and manipulations of these once-acoustic tones. This generalisation of samplers for instrument recreations and synthesisers for new sounds from scratch is especially broad, with each synthesiser and sampler bringing different strengths and weaknesses. Several samplers are superb for mangling sounds into novel and fantastic tones; and several synthesisers are great at acoustically faithful tones. Key examples of these are explored in the following sections.

Regarding terminology, it is worth noting that the term 'synthesiser' can be used as an umbrella term that can also encompass samplers, particularly in the context of plugins. The term 'sampler', however, does not encompass synthesisers. For clarity, the terms are kept separate in this book.

Synthesisers and samplers have become so commonplace that they have long since transcended the status of new instruments and have become embedded in most of the computers we interact with daily. Although the main types of synthesisers and samplers have been established for several decades, new ways to combine, use, and market these different devices are constantly being developed, ensuring a regular stream of new plugins and hardware devices.

The trend of devices that combine synthesis and sampling elements, otherwise known as 'hybrid' devices, has continued to the point where it is often hard to say what type of synthesiser or sampler is being used and at what point the sampling ends and the synthesis starts. It is also true that several vital components and typical workflows involved in electronic sound creation and manipulation are common to most kinds of samplers or synthesisers. In particular, it is worth noting that the sound manipulation elements for both samplers and synthesisers are essentially the same beyond signal generation. Because of this, sampling and synthesis techniques are both discussed in this chapter. Where separation is necessary, the letters 'a' and 'b' are used in the section titles: 'a' for synthesis and 'b' for sampling. Also, this book uses the full name 'synthesiser' rather than the abbreviation 'synth'.

3.1 KEY COMPONENTS

A model and starting point for how many synthesisers and samplers (as well as many acoustic instruments) work is the generation, manipulation, and modulation sources model, shown in Figure 3.1.

This model is a two-stage sound generation and manipulation arrangement, with modulation sources that can vary the parameters of either the generation or manipulation stages.

The generation stage is where the initial signal is generated, shown in the top left of Figure 3.1. For synthesisers, this is typically one or several oscillators, illustrated by a sawtooth (oscillators are discussed in Section 3.2a). For samplers, this is a sample editor, illustrated by an audio waveform (sample editors are discussed in Section 3.2b). The initial signal is then sent to the manipulation stage, which significantly alters the signal's tone, typically using several filters and an amplifier.

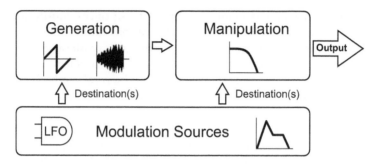

FIGURE 3.1

Generation, manipulation, and modulation sources model of synthesis and sampling

The manipulation stage is shown in the top right of Figure 3.1 (illustrated by a symbol for a low pass filter). Some synthesisers and samplers offer sound manipulation elements such as effects (the term 'effect' is used to describe both effects and parameters from this point), which in turn may be modulated. However, effects are not standard manipulation elements for synthesisers or samplers, and are covered in Chapter 2, so they are not discussed in this chapter. Performance, chance, reactive, and interactive adjustment of effect parameters is discussed further in Chapter 5.

In synthesis, the generation and manipulation stages alone typically create an unchanging, static tone. The static tone is made more interesting by modulating parameters over time. 'Modulation' in synthesis and sampling means altering the parameters of the signal generators or manipulation elements over time in a mechanistic fashion (rather than human expression). The mechanisms that cause the modulation are referred to as 'modulation sources'. Standard modulation sources include low-frequency oscillators (LFOs) and envelopes, and their respective symbols are used in the lower part of Figure 3.1 (discussed further in Section 3.4). The modulated parameters are referred to as modulation destinations – for example, the amplifier level, which typically has its own envelope (the modulation source) to control the tone level over time.

As well as the source and destination elements of modulation, there may also be a modifier (or 'via') element – an absolute value typically set aside for performance variation using a modulation wheel or another MIDI continuous controller (CC) message.

Some modulation arrangements are 'hardwired' as part of the design of a synthesiser (e.g., the amplifier or amp envelope). Other plugins use drag-and-drop or modulation matrixes. A modulation matrix is either a grid or a table allowing the connection of sources, modifiers, and destinations, and helps to illustrate the elements of modulation in synthesis. The matrix tab of the synthesiser plugin Wavetable in Ableton Live is a suitable example of a user-friendly modulation matrix and is shown in Figure 3.2a. As can be seen, the sources are displayed horizontally, and the destinations are shown vertically with the amount at the point that they join.

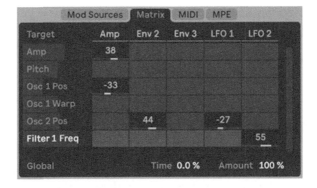

FIGURE 3.2

Matrix tab from the Wavetable plugin by Ableton

FIGURE 3.3

The Matrix from the Padshop 2 plugin by Steinberg

A more expansive matrix in the form of a menu-driven table can be seen in the synthesiser plugin Padshop 2 by Steinberg, shown in Figure 3.3. The source, offset, modifier, depth, and destination settings can be seen as options in the more standard column arrangement.

Sampling overview: In sampling, the generation stage causes the playback of a sample, and the manipulation stage offers the possibilities of sound manipulation tools – which, as mentioned in the previous section, are likely to be similar to the sound manipulation tools used in synthesis. However, as mentioned previously, while many samplers can apply modulation, it is not necessarily an essential element unless the sample itself is overly static or the sampler is used for specific sound design or creative tonal manipulation tasks. One of the primary uses of a sampler is to recreate the sound of a sampled instrument authentically, so a sampler will typically not use as much sound manipulation and modulation as a synthesiser. However, for approaches where the intention is to manipulate the sound of a sample creatively, the modulation capabilities of many samplers often match the flexibility and power of the most intricate synthesisers, allowing the sample to be completely transformed.

Samplers are particularly useful for automating repetitive manual audio editing techniques (see Section 2.3.1). For example, a small piece of audio (a sample), such as a kick drum, could be edited, copied, and placed repeatedly using manual editing techniques. This manual approach is more of a brute-force option; and while this does take longer, it gives more precise control over individual editing and rendering options. In contrast, using a sampler is likely far more efficient at the same job and opens up a wide range of modulation possibilities.

The elements that comprise these critical components of samplers and synthesisers, and how they create sound, are discussed in the rest of this chapter.

3.2A OSCILLATORS

As mentioned in Section 3.3.1, one or more oscillators generate the initial signal for synthesisers. An oscillator is an electronic device that creates a repetitive signal. In the signal generation stage of a synthesiser, this signal will be in the audio range of between 20 hertz (Hz) and 20 kilohertz (kHz). The waveform of the oscillator will determine the initial timbre of each of these signals. The specifics of each main waveform type are discussed in Section 3.6 in the context of their synthesis type.

3.2B SAMPLE EDITOR

As discussed in Section 3.1, the sample editor is a sampler's initial sound generation stage. As a sample is a section of recorded audio, the main features of a sample editor are typically setting the start and end points and the loop areas. The two main loop options are either sustain or release loops: sustain for when the note is held down and release for when the note is released (note envelopes are discussed in more detail in Section 3.4.1).

Figure 3.4 shows the sample tab of the Ableton Live plugin Simpler (a 'simple' sampler) with an electric piano sample. The sample's waveform is displayed with the start and end points of the sample at the small triangle at the top of the waveform view and the sustain loop shown.

Unlike the technique of sample editing in the context of a digital audio workstation (DAW) discussed in Section 2.3.1, the settings of a sample editor in a sampler are likely to be non-destructive (i.e., the sample file itself is not changed).

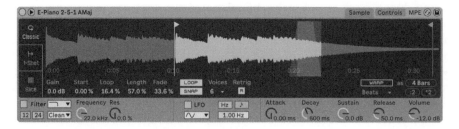

FIGURE 3.4
Sample tab from the Simpler plugin by Ableton

3.3 FILTERS

Filters are the main elements of the manipulation stage shown in Figure 3.1. As discussed in Section 2.2.3.1, filters adjust the frequency content of a signal. In the recording studio workflow context, corrective and creative sound manipulation techniques are typically added to the sounds that make up the mix. In synthesis and sampling, however, these techniques are likely to form an integral part of sound creation.

Filters and their controls are discussed in Section 2.1.1.1, and their symbols are shown in Figure 2.3. These symbols are shown in a sampling context in Figure 3.5.

Low pass filters are the typical default filter type for synthesisers and the main 'subtractive' element of synthesis because the tones created in the generation stage are likely to have an abundance of high frequencies. Accordingly, any cutoff frequency modulation (FM) tends to be in a frequency range sensitive to the human ear. As with the low pass filters in equalisers (EQs) used in the recording studio, the controls of low pass filters used in synthesis are cutoff frequency and slope. In synthesis, a resonance control is also made available.

As described in Chapter 2, the frequency cutoff is the point at which the filter reduces frequency content by 3 decibels (dB). The slope control sets how quickly the filter reduces frequencies over the frequency cutoff. Unlike the Q control of a bandpass filter – a continuous gradient with a decimal point typically between 0 and 10 – the slope control usually increases in steps of -6 dB per octave. A first-order (or one-pole) low pass filter will reduce the frequencies by 6 dB above the cutoff for every octave (i.e., for every doubling of the frequency); a second or two-pole low pass filter will reduce the frequencies by 12 dB/octave above the cutoff; while the maximum filter slope is typically a fourth-order, four-pole -24 dB/octave response, which is likely to be the most aggressive slope setting.

The resonance control boosts the frequencies just below the cutoff frequency, creating a peak that accentuates the sound of the filter and gives the classic analogue filter sound, which is often swept to dramatic effect. Three frequency responses of a two-pole ladder design low pass filter with varying resonances can be seen in the frequency response graph of the filter in the drum sampler Battery 4 by Native Instruments, shown in Figure 3.6.

A sufficiently high resonance setting will also allow the filter to self-resonate, creating a distinct pitch at the resonance frequency. It is also worth noting that the channel EQ in the DAW Logic X does offer resonance controls for low and high pass filters (in contrast to the more traditional EQ III in Pro Tools discussed

⌐ Low pass filter (LPF)
⌐ High pass filter (HPF)
⌒ Band pass filter (BPF)
⌣ Notch filter

FIGURE 3.5

Filter symbols from the Simpler plugin by Ableton accompanied by their names

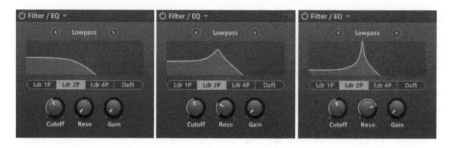

FIGURE 3.6

Three different resonance settings from the Filter/EQ module of the Battery 4 plugin by Native Instruments

in Section 2.1.1). A resonance control in EQ is a noteworthy example of synthesis techniques cross-fertilising into more general mixing techniques.

As with the filters that make up the EQs discussed in Chapter 2, some filters act transparently, with a clean sound and little character (other than the changes made to the specific frequencies that are boosted or cut); while other filters have a particular sound, imparting a character of their own.

The way that a filter changes the frequency content of a sound is determined by its particular design. Going into the specific details of filter design is outside the scope of this book, but notable filter design types are referred to in the following descriptions. The character that a particular filter gives is likely to be a certain kind of saturation or distortion that several factors will determine:

- the filter design (and its components if it is hardware or modelled on hardware)
- how much signal is present at its input (the drive control), and
- how much of its output can be fed back into its input (the feedback control).

Each different filter will react differently to varying levels of drive or feedback. The filter types are named after the modelled synthesiser type without subjective sonic descriptors. As with many creative pursuits, language is often woefully inadequate at conveying abstract characteristics, which are often quite personal.

As the saying goes: the map is not the territory. In this case, the linguistic map is not the sonic territory. (Map and territory relationships in the context of systems for liveness are explored further in the author's forthcoming academic book on liveness in modern music performance.)

The Ableton Live Simpler plugin offers a helpful selection of filter design types that are shown in Figure 3.7 and are described below:

- Clean: The most transparent filter setting, the same as the filters in the Ableton Live effect EQ Eight.
- OSR: A state variable filter design probably modelled on the British OSCar synthesiser.
- MS2: A Sallen Key filter design probably modelled on the Korg MS20 synthesiser.

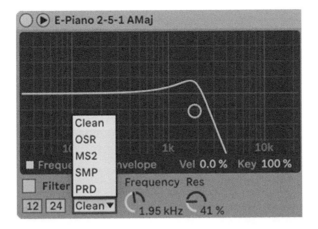

FIGURE 3.7
Filter types available in the filter controls section of the Simpler plugin by Ableton

- SMP: A custom filter design not based on any particular hardware, but a hybrid between the MS2 and PRD designs.
- PRD: A ladder filter design probably modelled on the Moog Prodigy synthesiser, which the band The Prodigy named themselves after.

3.4 MODULATION

The modulation sources shown in Figure 3.1 are devices that cause alterations to destination parameters (e.g., filter cutoff) over time. As discussed in Section 3.1, the main types of modulation sources are either LFOs or envelopes. Additionally, step sequencers may also be available in some synthesisers or samplers. These elements are described in the following sections.

3.4.1 Envelopes

Envelopes control signals with a limited number of stages triggered by the playing of a note. This signal is used to raise and lower a parameter of either the tone generation or manipulation stage.

A basic envelope has four sections: attack, decay, sustain, and release (ADSR). These sections are shown in Figure 3.8. The attack stage is the time that it takes the control signal to go from zero to its maximum level and, in a MIDI system, is triggered by a note on message (discussed in Section 4.2.1). The decay stage occurs immediately after the attack and is the time it takes to transition from the maximum level to the level specified by the sustain control (so if the sustain is set to maximum – that is, 100% – there is no decay stage). The sustain control is a *level* control rather than a *time* control, because sustain *time* is set by how long the player sustains the note (typically by holding down a piano key). Finally, the release control is the time that the control signal takes to go from

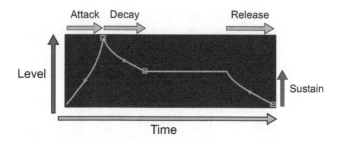

FIGURE 3.8
Annotated ADSR envelope from the Wavetable plugin by Ableton

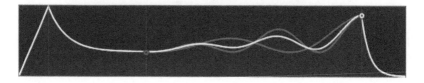

FIGURE 3.9
An envelope with looping features from the Massive plugin by Native Instruments

the sustain level down to zero and is triggered by a note's release – for example, taking a finger off a piano key.

The ADSR envelope is the standard type of envelope. There are several alterations to this standard and many more complex variations, such as the novel envelopes in the synthesiser Massive by Native Instruments, an example of which is shown in Figure 3.9. These envelopes allow a sustain loop with various selectable shapes, which can be morphed between, resulting in a kind of modulate-able LFO within the sustain section of the envelope.

3.4.2 Low-frequency oscillators

LFOs are oscillators that create a repetitive signal with a rate typically lower than the lowest frequency of human hearing, 20Hz. The term 'rate' is generally used instead of 'frequency' to differentiate the LFO from an audio frequency oscillator. As with envelopes, this signal is used to alter the destination parameter. However, an LFO does this repeatedly rather than an envelope's one-shot process (per note).

The shape of the waveform sets how the alteration occurs. For example, a sine wave will smoothly transition from minimum to maximum, whereas a square wave will switch almost instantly from minimum to maximum. A triangle wave will sound like a more angular sine wave, and a sawtooth wave will drop from maximum to minimum to return quickly to the maximum. Other wave shapes are available, adding a range of different characteristics. However, the more complex the waveform, in general, the less clear the modulation becomes, especially at higher rates. Figure 3.10 shows the waveform selector section of the

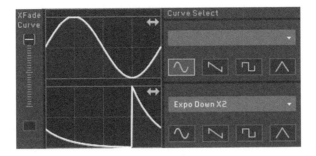

FIGURE 3.10

The LFO selectors from the Massive plugin by Native Instruments

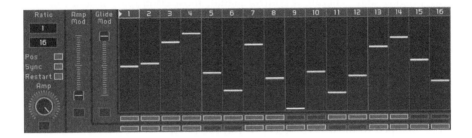

FIGURE 3.11

The step sequencer from the Massive plugin by Native Instruments

synthesiser Massive by Native Instruments (the term 'curve' rather than waveform is used in the LFO section of Massive). The upper waveform is a sine wave, taken from the four standard analogue type buttons of sine, sawtooth, square and triangle. The lower waveform is an exponential drop, taken from a large menu of over 30 more complex waveforms. As is often the case, the phase of the LFO waveforms can be offset. In Figure 3.10, the lower waveform's phase is shifted 270°.

3.4.3 Step sequencers

A less common form of modulation is the step sequencer, which steps through a series of set values that tend to give a choppy and rhythmic pattern type of modulation. The step sequencer acts as a complex form of synchronised LFO, where the segments of its control signal are highly customisable. There are typically up to 32 steps in a bar. Figure 3.11 shows a 16-segment step sequencer set to step through in 1/16th notes (so that it lasts a bar).

3.4.4 External modulation sources

In regular use, the modulation sources discussed in Sections 3.4.1 to 3.4.3 are an integral part of an electronic instrument; they are internally assigned, routed, or virtually patched inside that plugin or hardware unit. Internal modulation allows an optimal workflow for the instrument – be it a quick, pre-routed subtractive synthesiser or an open, deeply editable hybrid synthesiser. It also means that the design of sources and destinations can be matched to the needs of that instrument, giving smooth transitions and practical ranges.

There are, however, situations where it would be helpful to modulate a particular parameter, but the device design does not allow modulation by internal sources. Also, it may be beneficial to have several parameters of different devices modulated by a single common source. In these situations, it may be possible to modulate these parameters with sources that are external to the device or devices.

In most DAWs or plugin hosts (see Section 4.4), it is possible to use additional devices to work as an external modulation source. These devices are typically plugins that generate a control signal, most commonly a series of MIDI CC messages (see Section 4.2). These control messages can be mapped to any destination parameter that the DAW allows, typically using MIDI mapping or MIDI learn modes.

Examples of third-party external modulation source generators include the LFO Tool from Xfer records and the MidiShaper by Cableguys, both of which can run in various DAWs. In Ableton Live, the Max for Live devices LFO, and Envelope Follower (shown in Figure 3.12) are external modulation sources that can be mapped to most parameters within the DAW.

An envelope follower is a device that creates a control signal from incoming audio and then uses that control signal to change a parameter, such as the cutoff frequency on a filter (a built-in feature of the Auto Filter plugin). The Max for Live envelope follower device for Ableton Live is shown in Figure 3.12.

It is also worth noting that most analogue modular equipment is designed for modules to modulate and be modulate-able. The standard for modular equipment is control voltage (CV), which is a 1v per octave standard. CV affords a massive variety of modulation combinations depending on how the modules are patched together (modular synthesis is discussed in Section 3.6a.8).

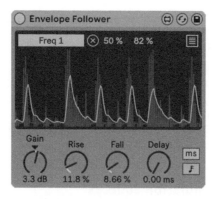

FIGURE 3.12

Ableton Live Envelope Follower Max for Live plugin

3.4.5 Modulation versus automation

Modulation and automation techniques are both system methods for altering sound manipulation parameters.

In modulation, the parameter is altered mechanistically by one of the modulation sources discussed in Sections 3.4.1 to 3.4.3. Modulation is discussed in this chapter in the context of synthesis and sampling as a part of instrumental tone sound creation. However, modulation is also a factor in broader sound manipulation scenarios (e.g., the modulation effects discussed in Section 2.2.3).

Automation, in contrast, causes a parameter to be altered by an automation graph, which operates relative to the timeline of the DAW playback. An example of volume automation for a vocal part in Pro Tools is shown in Figure 3.13. The Bar|Beats and Min:Secs position of the timeline can be seen at the top of the figure. As shown, the volume drops (to 0.2 dB) in the middle of bar 92, then peaks briefly (at 7.1 dB) at the beginning of bar 93. Automation is typically used as a part of the mixing process rather than instrument tone/sound creation, as is discussed in Section 2.1.5.

Because automation is tied to the timeline of the DAW playback, automation will only alter the parameter when the transport of a DAW is running and can only be written along the timeline of the DAW. Modulation, however, is not directly tied to the timeline of a DAW, so it does not necessarily need the transport of a DAW to run. Modulation can be set to operate every time the sound is played in re-trigger mode or to operate freely, independent of the sound being played. Consequently, the modulation source (e.g., LFO or envelope) only needs to be set once, and the operation will be repeated.

Because of these factors, automation is generally better suited to precise, non-repetitive parameter alteration on playback – the most typical example being the automation of a track's volume fader to even out performance dynamics. Modulation, on the other hand, is generally better suited to parameter alterations that repeat according to the source and modifier settings.

In traditional DAWs, it is typically the case that modulation occurs at the instrument level and automation occurs at the DAW level. DAWs that have clip launching as an integral part of their workflow (discussed in Section 4.4.2) are exempt from this general rule. This exception is because a system of clip

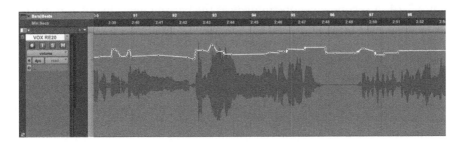

FIGURE 3.13

Volume automation in Pro Tools

FIGURE 3.14

Automation envelope in session view of Ableton Live showing the breakpoint value (dB)

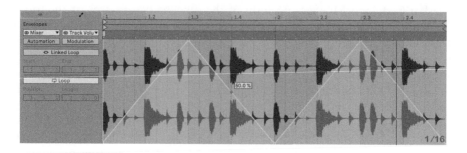

FIGURE 3.15

'Modulation' envelope in session view of Ableton Live showing the breakpoint value (%)

launching (e.g., the session view of Ableton Live) works similarly to a kind of macro-sampler: part DAW and part sampler. Put another way, a clip-launching DAW can manage the triggering of samples and clips in various non-linear methods. For example, in Ableton Live, a clip in the session view can have several automation envelopes, which are set up like standard automation. However, the envelopes repeat every time the clip is launched, like a modulation source; they are not directly linked to the timeline.

To further complicate terminology, clips in the session view can be given two types of envelopes. The first sets the absolute parameter level (e.g., in Hz for a filter cutoff), is called the automation envelope, and is shown in Figure 3.14. A second envelope influences the percentage value of automation. It is called – with a high potential for confusion – the modulation envelope, and is shown in Figure 3.15. (For comparison, in the clip-launching DAW Bitwig, the equivalent envelope is called 'relative automation', and can be additive or multiplicative.)

If that clip is placed in the arrangement view, the automation envelope becomes an automation graph. The automation graph is then shown in the track and can be edited differently for every concurrent instance of the clip. Modulation, however, stays as the sole clip envelope. Figure 3.16 shows this same clip pasted into the arrangement view.

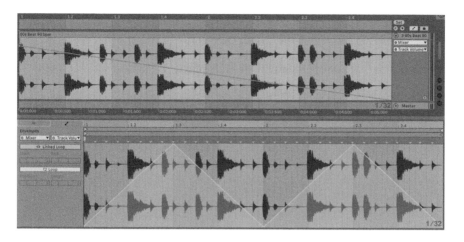

FIGURE 3.16

Automation and modulation viewed in the arrangement view of Ableton Live

3.5 CATEGORIES OF SYNTHESIS PATCHES

In general use, synthesiser patches (i.e., presets) can be separated into three categories: bass, pad, and lead patches. These categories can help to search for a patch for a particular use, but the categories are not hard and fast rules for patch creation, and there can be significant overlap between the categories. An additional quality of a synth patch, which can also be used to categorise it, is whether it is monophonic or polyphonic. These terms are explained in the following paragraphs.

Bass: As the name suggests, bass patches are sounds with predominantly low frequencies – that is, notes under middle C, C4 or 262Hz (this assumes a bass/treble split without considering middle frequencies, generally in the range of 200–3000Hz). While bass patches can usually play notes over middle C, these can potentially sound harsh or thin.

Pad: A synthesiser pad is typically a highly textural and atmospheric patch with a slow attack and release, lending itself to an underpinning accompaniment useful for 'padding'.

Lead: A synthesiser lead is typically the sonic opposite of a pad: a patch with a clear and distinctive sound, a sharp attack, and a relatively quick release. A lead patch is likely to be better suited to a solo part.

Monophonic/polyphonic: The distinction between a monophonic synthesiser and a polyphonic synthesiser concerns the technical capability of a synthesiser. A monophonic synthesiser (commonly referred to as a 'mono synth', but the full terms are used here for consistency) can only play one voice (or note played) at a time. In contrast, a polyphonic synthesiser (often referred to as a 'poly synth') can play more than one voice at a time. Polyphonic synthesisers will have a voice limit or polyphonic limit that describes how many voices can be played at once.

While not necessarily a direct factor of the synthesiser's timbre, a monophonic synthesiser is likely well suited to bass or lead patches. On the other hand, polyphonic synths are well suited to pads, leads and bass parts, and are likely to have a notably different timbre to similar sounds played on a monophonic synth.

3.6A SYNTHESIS

As with many other technologies discussed in this book, a generally chronological approach is adopted in ordering different synthesisers. As the first types of synthesisers were subtractive, they set many standard workflows and design techniques across other types of synthesis.

3.6a.1 Subtractive synthesis

Subtractive synthesis is the earliest type of commercially available synthesis. Subtractive synthesis is the typical type meant by analogue synthesis, as most other main synthesis types require digital technology. As a general description, the sonic characteristic of a typical early subtractive synthesiser (often abbreviated to 'sub synth'), such as the Minimoog, is a warm, lush tone but with the potential to sound quite muddy. Subtractive synthesis tones can be heard in much of the music produced in the 1960s and 1970s – for example, the bass sound of Stevie Wonder's 'Livin' for the City' or the lead sound of the solo in Emerson Lake and Palmer's song 'Lucky Man'.

As discussed in Section 3.2a, the first stage of a subtractive synthesiser is a set of oscillators that use only a limited number of waveforms. Other than noise, the typical waveforms are sine, sawtooth, square/rectangle and triangle waves. Each of these waveforms has a distinctive sound, and importantly, all except the sinewave are complex waveforms – they contain frequencies above the fundamental harmonics. The more complex the waveform, the more frequencies are present (typically more high frequencies). Accordingly, there are many sonic elements to manipulate, which lends to subtractive techniques, where one or more filters are used to remove various frequencies, allowing a sort of sonic sculpting.

The timbral characteristic of each waveform can be described by the harmonic content and with more general descriptive terms. The oscillator view in the Ableton Live synthesiser Operator allows a handy visual indication of both the waveform shape and the content of the first 16 harmonics displayed in idealised bars. Figures 3.17 to 3.21 show the waveform (in the bottom-right corner of each screenshot) and harmonic content (the blue bars) of the Ableton Live synthesiser Operator for the four basic subtractive waveforms and a waveform determined by a user-specified harmonic content.

Sine wave: As mentioned previously, an ideal sine wave (shape and harmonic content shown in Figure 3.17) only has energy at its pitch frequency (also called its fundamental frequency), so it has no overtones. In reality, it is worth noting that there are likely to be some small amounts of overtones due to imperfections or nonlinearity. A sine wave is not particularly useful in a subtractive synthesiser as there are no overtones to subtract (except when used to underpin a patch at the fundamental frequency or an octave below).

FIGURE 3.17

Waveform and harmonic content of a sine wave from Operator plugin by Ableton

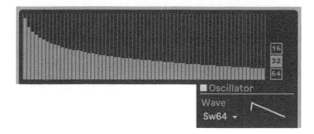

FIGURE 3.18

Waveform and harmonic content of a sawtooth wave from Operator plugin by Ableton

Sawtooth wave: A sawtooth wave (shape and harmonic content shown in Figure 3.18) has a harsh, buzzy tone; but compared to a square wave, it is richer in timbre as it contains both even and odd-order harmonics. Because of the rich harmonic content, a sawtooth wave is often a good starting point for a typical subtractive synthesiser tone. It is worth noting that a sawtooth wave is a linear transition from maximum to minimum across the wavelength (top left to bottom right). The opposite shape is called a ramp – a wave that is a linear transition from minimum to maximum across the wavelength (bottom left to top right). Both a sawtooth and ramp waveform have the same harmonic content.

Square wave: A square wave (shape and harmonic content shown in Figure 3.19) also has a harsh, buzzy tone. Compared to a sawtooth wave, it has a hollow timbre as it only contains odd-order harmonics – that is the fundamental frequency multiplied by the odd whole numbers (3, 5, 7, and so on).

Triangle wave: A triangle wave (shape and harmonic content shown in Figure 3.20) also has the hollow timbre of a square wave, as it contains only odd-order harmonics, but these harmonics have significantly less power than a square wave, so its tone sounds far more mellow.

User-defined wave: The harmonic content of a user-defined wave (shape and first 16 harmonics shown in Figure 3.21) is like a sawtooth. However, the even-order harmonics have been boosted, and the odd-order harmonics have been lowered; the resulting simplified waveform resembles a lopsided sinewave.

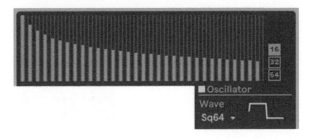

FIGURE 3.19

Waveform and harmonic content of a square wave from Operator plugin by Ableton

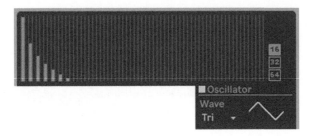

FIGURE 3.20

Waveform and harmonic content of a triangular wave from Operator plugin by Ableton

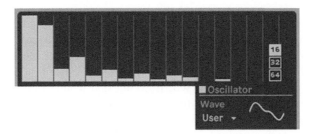

FIGURE 3.21

Waveform and harmonic content for a waveform determined by user-specified harmonic content from Operator plugin by Ableton

It is worth noting that the transitions of a square wave are represented by the vertical lines indicating instant transition from minimum to maximum and maximum to minimum. (This is the same for the vertical lines of a sawtooth and ramp waveform.) While a digital waveform can be at a minimum at its last sample and a maximum at its first sample, a loudspeaker – or any other analogue system such as voltage or magnetic charge – needs time to travel from its minimum to its maximum point. An instant transition would take infinite energy and create an endless number of high-frequency harmonics, so it is impossible. However, vertical lines are handy to form an ideal graphical representation of basic wave shapes. When using graphic waveforms, a similar and general rule related to the point about vertical lines is worth noting: the more vertical the lines are, and the more jagged a waveform is, the more high frequencies are present.

3.6a.1.1 Hard sync

A function in many subtractive synthesisers that allows more variation in the complexity of the waveform is hard sync. Hard sync is when one primary oscillator forces the resetting of the waveform of another secondary oscillator for each waveform cycle. This resetting creates a complex waveform when the frequency of the primary oscillator is lower than that of the secondary. If the frequency of the secondary oscillator is swept, then a characteristic tearing sound is produced.

3.6a.2 FM synthesis

Access to the digital technology of the late 1970s brought a type of synthesis fundamentally different from the warm tones of subtractive synthesis. FM synthesis has a characteristic clean and clear sound, with the potential to sound quite brittle. FM synthesis tones can be heard in much of the music produced in the 1980s – for example, the punchy bass sound of the A-ha song 'Take on Me'; and the glassy, ambient tones of the Brian Eno piece 'An Ending (Ascent)'.

FM synthesis makes use of the radio technology of the same name. FM radio transmissions work by taking a very high-frequency signal (e.g., 100 MHz), called the carrier, and modulating the carrier's frequency with an audio signal (e.g., speech or music with content between 20–20000Hz), called the modulator. The combination of these signals can carry long distances before the signals are received by an FM radio, which removes the carrier using a low pass filter and plays back the audio signal.

The tremolo effect, described in Section 2.2.3.3, works by modulating the *amplitude* of an audio signal using an LFO. In Figure 2.16, the modulation causing tremolo can be seen in the amplitude, or height, of the waveform. If an audio signal has its *frequency* modulated by an LFO, then this effect is vibrato. The change in pitch will be an audible variation, which, if applied aggressively, can sound like a siren. A visual representation of FM with an LFO is shown in Figure 3.22, where the modulation can be seen in the lengthening and shortening of the output signal's wavelength as the frequency is decreased and increased by the LFO.

LFO   modulates   frequency

FIGURE 3.22
Diagram showing frequency modulation/vibrato

FM synthesis works in the same way as FM radio and vibrato; however, both the carrier and modulator are audio rate signals, and traditionally are both sine waves. When the modulator and carrier are both at an audio rate (20 Hz–20 kHz), instead of the frequency variation being audible (as with vibrato), a particular type of distortion occurs as additional frequencies called sidebands are created. The more the modulator is set to modulate the carrier, the stronger the sidebands will be, and the more there will be. The frequencies of these sidebands depend on the relationship between the carrier's and the modulator's frequency. If the modulator and carrier frequencies are whole number multiples of each other (i.e., if they are harmonically related), then the sidebands will be harmonically related, giving the sound a consonant timbre. This relationship is often described as a ratio, so if one of the sides of the ratio is 1, the relationship between them will be harmonic if the other side of the ratio is a whole number. For example, if the modulator is 220 Hz and the carrier is 440 Hz, then the ratio of the frequencies will be 1:2, and the sidebands will be harmonically related.

If the frequencies of the modulator and carrier are not whole number multiples of each other, then the sidebands will not be harmonically related and are likely to sound more dissonant the further they move away from whole number multiples. Particularly dissonant tones are likely to sound clanging or harsh.

More complex timbres can be created using multiple modulators and carrier combinations. The strength of each signal over time is critical to how the tone of each note evolves, so FM synthesisers typically combine each oscillator with an envelope (rather than subtractive synthesisers, which are more likely to have envelopes applied to the overall tone). Rather than referring to each separately, the pairing of the oscillator and envelope is called an operator.

The arrangement of operators as modulators and carriers can become very complicated, so they are typically represented by an algorithm that specifies how each operator modulates the other. For example, Figure 3.23 shows a synthesiser tone with four operators arranged in the algorithm that is enlarged in

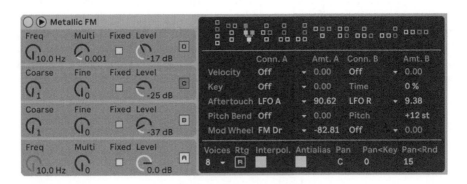

FIGURE 3.23

Operators and algorithms sections of the Ableton Live Operator plugin

Figure 3.24. The algorithm shows that operator C is the modulator for operator B, which is the carrier. The resulting tone combines with operator D, which in turn modulates the frequency of the final carrier, operator A.

The oscillators of early FM synthesisers, such as the very popular Yamaha DX7, had only sine waves as waveforms. Sine waves are suited to FM synthesis because when FM is applied in several stages, the waveforms quickly become complex, and an extensive range of tones can be created. The DX7 was capable of extensive user patching; however, the interface and procedure of FM were so complicated that few musicians went beyond the 32 factory presets. Specifically, preset 11 (Electric Piano) and preset 15 (Bass 1) can be heard in a surprising number of popular music songs in the 1980s – for example, 'Hard Habit to Break' by Chicago and 'Bad' by Michael Jackson.

It is worth noting that the traditional FM synthesis process is almost the polar opposite of subtractive synthesisers – that is, the degree of tonal complexity is typically created by the varying amount of FM and the cumulative arrangement of modulators and carriers. In subtractive synthesis, tonal variation is generally controlled by varying the parameters of filters to take away tonal elements.

This workflow results in a significant variation of the generation, manipulation, and modulation sources model shown in Figure 3.1, where the operator algorithm interweaves the often otherwise discrete stages of generation, manipulation, and modulation sources.

Modern synthesisers – even ones that focus on FM (e.g., FM8 by Native Instruments) – are unlikely to be limited solely to a traditional FM process unless they are modelled directly on vintage synths (e.g., Dexxed by Digital Suburban). The oscillators in modern FM synthesisers often offer a broad range of waveforms rather than just sine waves, making the resulting tones even more tonally complex. Modern FM synthesisers are also likely to include filters and modulation capabilities, allowing the blending of FM and subtractive methods.

As a side note, audio rate amplitude modulation (AM) is a far less popular technique for sound manipulation than FM as its tones are particularly metallic and harsh. However, the ring modulation effect is a dramatic variation on audio rate AM as it is purely the side bands of AM (with no original signal present).

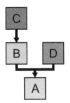

FIGURE 3.24

Expanded flow diagram of the third Operator algorithm of the Ableton Live Operator plugin

3.6a.3 Wavetable synthesis

Wavetable synthesis uses look-up tables, or wavetables, as its primary tone generation stage. A wavetable is a series of tiny samples, often as short as a single waveform cycle. The samples or waveforms in the wavetable can be smoothly moved through by a modulation source so that the tone can have a dramatically shifting timbre.

A good illustration of a basic wavetable is the Wavetable VCO (WT VCO) module in the virtual Eurorack studio, VCV Rack, made by VCV. Figure 3.25 shows the top sections of three WT VCO modules. Each module uses the Basic. wav wavetable. Basic.wav is a WAV file 4096 samples long, made up of the four basic waveforms – sine, triangle, ramp and square waves – each of which is 1024 samples long.

The module on the right of Figure 3.25 has a wavepoint setting of 4096 samples to display the whole wavetable. The wavepoint setting determines how many samples each waveform has so that, for instance, it cannot sweep the wavetable position, as it is a table of only one wave (albeit one made up of the four waveforms).

The modules in the centre and left of Figure 3.25 use the same wavetable but have wavepoint settings of 1024 so that the wavetable position control can sweep across the four single-cycle waves. The module on the left has its wavetable position set to 33%, triggering a triangle wave. The module in the centre has its wavetable position set to 50%, halfway between the triangle and ramp waveforms, so the wave triggered is an even combination of the two waves.

An alternate way of visualising a wavetable can be seen in the synthesiser Wavetable by Ableton, which has a user interface that graphically illustrates a wavetable stacked vertically, as shown in Figure 3.26.

As with most wavetable synthesisers, the signal flow and method of operation are very similar to those of subtractive synthesisers, but they differ in the additional modulation capabilities that wavetables afford.

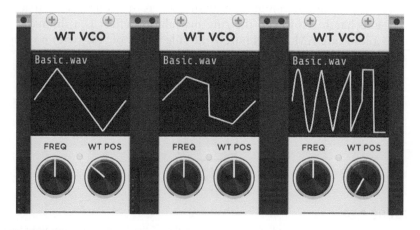

FIGURE 3.25

The top sections of three Wavetable VCO modules in VCV Rack 2

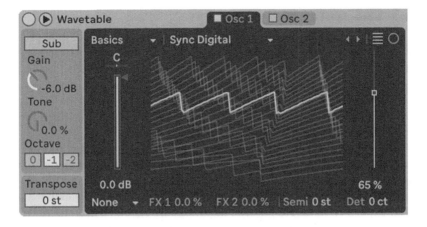

FIGURE 3.26

The oscillators section from the Wavetable plugin by Ableton

FIGURE 3.27

Ableton Live Granulator II Max for Live plugin

3.6a.4 Granular synthesis

Granular synthesis is a sample manipulation technique that slices an audio sample, often up to 30 seconds long, into many different 'grains'. These grains can then be played back and modulated in many ways. Granular synthesisers are particularly well suited to diffused and atmospheric sounding pads and glitchy rhythmic noise clouds. An example of a granular synthesiser plugin is the Max for Live Granulator II plugin shown in Figure 3.27. The sample is shown across the top of the graphic user interface (GUI), and the grains played at the instant the screenshot was taken are illustrated by the colourful lines on the sample.

Granular synthesis is an example of a modern sound manipulation technique that is as much sampling as it is synthesis. The Granulator II plugin in Figure 3.27 shows a sample display typical of a sampler at the top of the GUI. The other controls beneath the sample illustrate the settings for the granular processing, starting with the grain size and file position settings.

A modern variation on the granular synthesis method involves buffering an input signal to apply granular techniques as an audio effect – for example, the free/name your price plugin Emergence by Daniel Gergely.

3.6a.5 S&S synthesis

S&S synthesis (the 'S&S' standing for 'samples and synthesis', despite preceding the word 'synthesis') is an example of a technique that was short-lived but highly popular in the 1990s. An example of an S&S synthesiser is the best-selling synthesiser the Korg M1, responsible for the punchy and present dance piano sound (preset 01 'Piano16'), among other sounds characteristic of popular music in the 1990s. An appropriate example of this piano sound can be heard in the song 'Ride on Time' by Black Box.

As the name implies, S&S synthesis combines samples and synthesis modulation techniques. S&S synthesis was a solution to the problem that purely synthesis patches in the late 1970s and 1980s lacked a convincing attack when trying to replicate versions of acoustic instrument sounds. Computer memory was too expensive to create purely sample-based instruments, so S&S synthesis combined the strengths of synthesis and sampling within the boundaries of late 20th-century technology.

Typically, S&S synthesis used a sample for the attack and decay of the note; then, synthesis techniques were used to generate the sustain and release. The combination of these techniques resulted in an instrument capable of producing tones that were more realistic and exciting recreations of acoustic instruments.

3.6a.6 Additive synthesis

Additive synthesis is a technique where a complex timbre is created by adding individual sine waves. In additive synthesis, each harmonic or non-harmonic element of a sound's timbre is added by a different oscillator. Compared to subtractive, FM, and wavetable synthesisers, additive synthesisers are less common because they require many oscillators with many parameters. So many oscillators make additive synthesis less computationally efficient than most other types of synthesis, which means additive synthesisers are likely to require a complex control system to control the oscillator parameters.

Organs are an excellent example of an additive sound creation technique which pre-dates synthesisers. The drawbars of a Hammond organ set how much of each harmonic is added to each note played. Because Hammond organs generate tones using spinning tone wheels and pickups, which are inherently non-linear, the addition of these complex tones results in the Hammond organ's famous thick and textural sound. The broad distinction between synthesisers and organs is made not only by naming convention but also by the wide range of modulation capabilities discussed in Section 3.3.4 that typically distinguish the electronic instrument as a synthesiser rather than an organ.

3.6a.7 Hybrid synthesis

For hardware synthesisers (and other equipment), the term 'hybrid' can mean a combination of analogue and digital technology. However, many modern synthesisers, either hardware or software, are often a hybrid of several synthesiser types – for example, subtractive, FM, and wavetable. Consequently, it is rare for the information about a synthesiser to classify itself as just one type of synthesis (except for synthesisers modelled on vintage equipment).

A notable example of an expansive and inherently hybrid software synthesiser is the Synthmaster by KV331, which offers the synthesis types VA, additive, wavetable, phase modulation, FM, and physical modelling in a semi-modular setup.

3.6a.8 Modular synthesisers

As mentioned in Section 3.4.4, modular synthesisers are made up of separate modules, each being a synthesis or sound manipulation component (e.g., an oscillator, filter or mixer) that needs to be patched together using patch cables. Modular synthesisers were particularly common in the early days of analogue synthesisers and have seen a more recent increase in popularity for both hardware and software synthesisers. The Eurorack format is popular for hardware synthesisers, offering rack-mount devices that are easily interchangeable and often available as build kits.

Several audio programming environments have also developed higher-level modular synthesis styles of patching, such as Beep in MaxMSP and Blocks in Reaktor. The software instrument VCV Rack is a notable example of an open-source modular synthesis software environment that can be used as a standalone program or as a plugin.

Semi-modular synthesisers are devices capable of standalone synthesis in an initial hardwired configuration but can also be rearranged using patch cords.

3.6B CATEGORIES OF SAMPLES

This section discusses different sample categories and how they are used. (The practical aspects of sample production are discussed in Section 4.5.2.) Samples typically fall into two main categories: single-event samples and phrase samples. Within these categories, the musical qualities of being pitched or percussive will also influence how samples are used and manipulated.

3.6b.1 Single event samples

Single event samples are individual musical events over time. It is more common for single event samples to be just a single instrument voice played individually (not layered with other pitches or instruments), giving the sample patch much more flexibility. When multiple musical events are combined in single event samples, this may be several notes layered together and played simultaneously – for example, a string ensemble.

Single event samples are well suited for creating a sample patch that can be played back on a MIDI controller (typically a piano keyboard or set of pads, see Section 4.3.2). These patches can either recreate the authentic timbres of another instrument or allow further creative sound manipulation. As most instruments change their timbre and volume when played more loudly, more responsive dynamics can be created by using multiple samples captured at different dynamic levels and then triggering these different samples at different velocity levels. Creating many different samples to capture an instrument's wide timbral variety is called multisampling.

Single event samples can be further categorised into notes or hits, depending on whether they are pitched or percussive.

Notes: Notes are pitched sounds, which are sounds with one or more dominant and resonating fundamental pitches – for example, the individual notes of a piano (or the individually layered notes of a string ensemble).

Hits: Hits are percussive sounds with a wide range of frequencies, with no single dominating fundamental pitch – for example, the different types of sounds created by hitting a snare drum, cymbal or tabla. As the duration of a typical drum or cymbal hit is determined by the resonant characteristics of the instrument, most samplers allow a mode called 'one shot', which plays the sample until its end or unless a 'choke' instruction is received (typically a particular MIDI note message). The standard operation of a sampler is to play the sample until the note off/release message is received, which is effective for piano or string-like performance.

A well-produced sample collection will often have each hit or note saved as a separate audio file, organised into folders. In some cases, it is likely to be more computationally efficient to have all samples consolidated into one large file split into its parts by the sampler. Consequently, it is common to see percussion sample instruments with only one velocity layer and one long audio file, especially in the Ableton Live Sampler plugin (see Figure 3.26a).

While a well-produced sample collection will allow effective patch creation, there are also several standard preset files for sample arrangement, such as the Native Instruments Kontakt .nks files and Ableton Live .adg files.

3.6b.2 Phrase samples

Phrase samples are several consecutive musical notes or hits recorded over time to form a phrase – for example, one or several bars of a vocal melody, a chord progression on a piano, or a rhythm played on a drum kit. A phrase sample can also be referred to as a 'loop' if formatted to suit repeating. A good example is a drum loop, which is helpful for quickly creating a rhythmic accompaniment.

Famous loops that have been taken from commercially released songs include 'the funky drummer' – a rhythmic break in the James Brown song 'Funky Drummer (Part 1)' used in many hip hop releases; and the 'ahem break' – a distinctive beat from the Winstons song 'Ahem, Brother', which typically forms a central part of songs in the jungle dance sub-genre.

3.7B SAMPLERS

Most samplers follow the generation, manipulation, and modulation sources model shown in Figure 3.1 more closely than synthesisers. As well as this, as mentioned in Section 3.1, the main difference between synthesisers and samplers is that the sample editor is the generation stage (rather than one or more oscillators). Because of these two factors, there is one primary type of sampler: a general sample playback and manipulation engine typically referred to as a 'sampler'. The other kinds of samplers are either 'drum samplers', for the specific application of producing drums and beats, or 'sample players', for simplistic sample playback (discussed in Sections 3.3.6b.2 and 3.3.6b.3, respectively).

Examples of software samplers are the third-party plugin Kontakt by Native Instruments and the samplers included with Logic and Ableton Live, both of which are called Sampler. An example of a hardware sampler is the Roland SP-404 sampler, a standalone device triggered by a grid of 4×3 pads.

As well as the functions associated with the sample editor, many samplers also offer sound manipulation techniques related to synthesis (the modulation and filters discussed in Sections 3.3 and 3.4) and mixing (discussed in Chapter 2). However, as these techniques have been discussed previously, this section focuses on the functions that are typically exclusive to samplers – specifically the functions of the sample editor and the mapping elements of a sampler.

Sample editors provide non-destructive trimming and looping of samples – that is, they do not change the original audio file (the sample).

The software sampler Kontakt by Native Instruments is an example of a modern software sampler with a comprehensive feature set. Figure 3.28 is a cropped screenshot of a bass guitar ensemble patch in Kontakt, showing the sample editor (called the wave editor in Kontakt) at the bottom and the mapping editor at the top.

3.7b.1 Trimming

Trimming is the adjustment of the start and end point of the sample, typically adjustable using a marker on a graphic of the sample's waveform or by specifying the start and end point. The start and end markers (labelled S and E, respectively) of a sample in the wave editor of Kontakt can be seen at the bottom of Figure 3.28.

3.7b.2 Looping

The looping functions of a sample editor allow one or more sections of the sample to be repeated. There was very little computer memory in early samplers, so it was impossible to have sample-based instruments made from many long samples. To allow longer note lengths, the sustain section of a note (for a description of 'sustain' in this context, see Section 3.4.1) was looped – typically until the release message was received. While this allowed technically infinite note lengths, at best, short sustain sections are likely to sound static, and at worst, the cross-over points become apparent. Computer memory is now relatively inexpensive and accessible, so samples can be much longer if needed, capturing a sound's natural evolving complexity. Looping features are still available in most samplers; however, modern usage of looping is more likely to be for creative sound design purposes than to lengthen the playback of a sample (unless a particularly long playback is required).

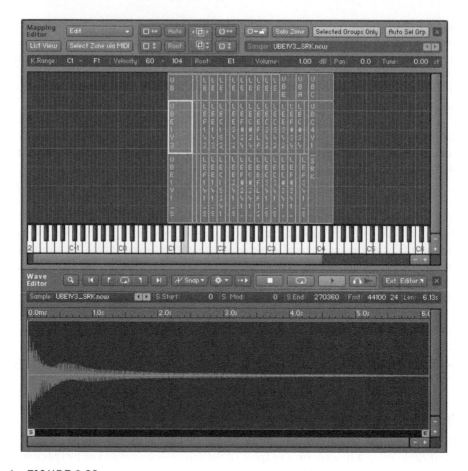

FIGURE 3.28

Mapping editor and wave editor from the Kontakt plugin by Native Instruments

A single loop of a sample in the wave editor of Kontakt can be seen in the yellow box at the bottom of Figure 3.28.

3.7b.3 Sample mapping

Samplers are typically triggered by MIDI note messages, so the playback of a sample is set by mapping it to one or more keys on a piano keyboard. As a sample is mapped across several different keys, the sample is either sped up or slowed down to match the pitch of the key pressed. (This is the same technique as the vari-speed pitch shifting discussed in Section 2.3.1.5.4.)

If the sample is of a pitched acoustic instrument, a general rule of thumb is that the pitch can be stretched approximately two semitones up or down until it sounds noticeably artificial.

As mentioned in Section 3.3.6b.1, different samples with different dynamics can be triggered at different velocity levels. Figure 3.28 shows the mapping editor of Kontakt, showing the notes across the horizontal axis and the velocity levels in the vertical axis. The selected sample in Figure 3.28 is spread over the notes C1–F1 and over the velocities 60–104. The patch shown in Figure 3.28 is typical of the approach needed to create a flexible and authentic-sounding sample-based plugin of a physical instrument (in this case an upright bass guitar), as there are many samples recorded at multiple velocity layers.

An alternative method of displaying sample mapping is showing samples in a list view, as depicted in Figures 3.29, 3.20 and 3.31, illustrating the key zone, the velocity zone, and the sample select editors of Ableton Live's Sampler respectively. The sample select editor is an alternate method for selecting different samples not tied to any specific element of MIDI messages. The sample select editor can be handy for switching out different groups of samples based on other data (e.g., in the reactive techniques discussed in Section 5.5.2). The patch shown in Figure 3.29 is typical of a sample-based plugin of a basic electronic instrument as there are far fewer samples needed, and shows that note pitches can be stretched further without obvious tonal degradation. Most notably, as there is no tonal difference between different dynamics, there only needs to be one velocity layer (as shown in Figure 3.30).

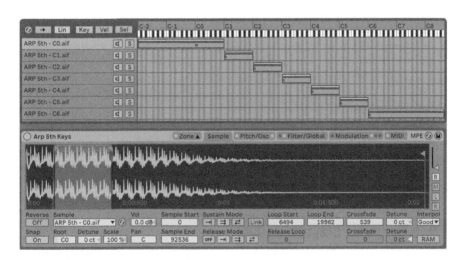

FIGURE 3.29

The key zone editor and sampler window from the Sampler plugin by Ableton

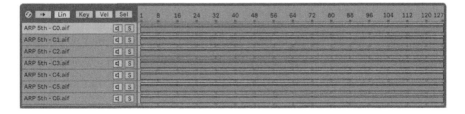

FIGURE 3.30

The velocity zone editor from the Sampler plugin by Ableton

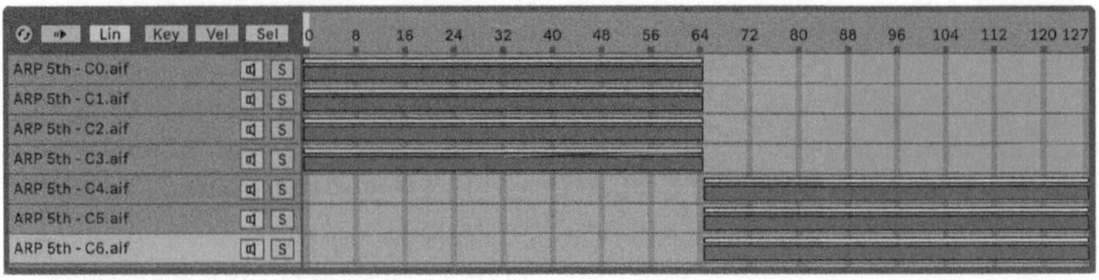

FIGURE 3.31
The sample select editor from the Sampler plugin by Ableton

3.7b.4 Standard playback mode

The standard playback mode of a sampler is to launch the sample when a note-on message is received and then fade out as per the release setting when the note-off is received. In this mode, the sampler controls the sample's playback by replicating the action of pressing and releasing a piano key. This method is generally satisfactory for the rudimentary triggering of most sample-based instruments imitating pitched acoustic instruments.

It is worth noting that for any instrument without a piano-type keyboard, this method is likely to lack the original way of playing. For example, playing a brass instrument or picking a guitar is very different from playing back these sounds using a MIDI controller. In most acoustic instruments, polyphonic notes will also interact in complex and unique ways depending on the physical elements of the instrument.

These are the main reasons why, however high quality the sample library of an instrument is, the sampler is unlikely to sound particularly authentic when triggered by MIDI 1.0 messages. However, sample-based instruments tend to be highly convenient, cost-effective, and consistent compared to their physical equivalents.

3.7b.5 One-shot playback mode

The alternative to standard playback mode (used chiefly for pitched instruments) is one-shot playback mode, mainly used for percussion instruments such as drums, cymbals, and mallet instruments. One-shot playback is helpful for sounds with a fixed duration that are required to be triggered initially and then be allowed to play all the way through. One-shot mode allows the resonance of the captured sample to sound out as it would with a physical instrument – for example, a sample of a cymbal crash or a snare drum. As with real cymbals, especially with a hi-hat choke, the playback of one-shot samples can be stopped with a MIDI note mapped to control the playback of one or more samples.

3.7b.6 Drum Samplers

As the name suggests, drum samplers are specifically designed for the playback and manipulation of drum samples that default to one-shot playback mode and offer manipulation tools specific to the environment.

Drum samplers typically fall into two categories of samplers designed for either beat manipulation, such as Battery by Native Instruments, or acoustic drum kit production, such as BFD3 by FXpansion.

3.7b.7 Sample players

Sample players are limited versions of samplers designed to offer playback functionality rather than the extensive sound manipulation functions of samplers discussed in Section 3.7b. Sample players are often less expensive than fully featured samplers – for example, Kontakt Player by Native Instruments, available at no cost (at the time of writing).

Tools of music technology

Chapters 1 to 3 focused on the underpinning concepts and techniques of music technology in live performance. In this chapter, the tools used for music technology are explored. The first two sections discuss the more figurative tools of practical music theory and Musical Instrument Digital Interface (MIDI) messages, and the remaining sections explore the more literal tools of the hardware and software used to make music.

As before, where there is an overlap between standalone hardware devices and software plugins, software plugins are explored here. Plugins are focused on because of the accessibility of computer-based systems such as digital audio workstations (DAWs) and the clearer and more consistent illustration of the techniques that graphical user interfaces offer.

Note naming and middle C conventions: When discussing Western music note pitches, it is convenient to use the MIDI system of naming the keys of a piano using standard musical letters and a number that increases at C for every octave. Also, the MIDI note display convention is to only use sharp symbols (#) for accidentals (i.e., black keys), so in the absence of musical context (i.e., key signatures), this convention is used here.

In terms of frequency and MIDI note number, middle C is set at 261.6 hertz (Hz) and MIDI note 60. However, there are two different conventions for the octave on which to place middle C. The traditional musical standard and the convention used by Roland place middle C at C4. The electronic music standard and the convention used by Yamaha place middle C at C3.

This book uses the Roland, middle C = C4 convention (as shown in Figure 1.3). Using this convention, the lowest key on a standard piano keyboard is A0, and the highest key on a standard piano keyboard is C8.

To give a specific example of an octave of notes starting from the typical lowest note on a piano keyboard, these are A0, A#0, B0, C1, C#1, D1, D#1, E1, F1, F#1, G1, G#1, A1, and so on. It is also worth noting that MIDI notes can also go negative, so the A an octave below A0 is A-1.

DOI: 10.4324/9781003370406-5

4.1 APPLIED MUSIC THEORY

This section introduces the practical components of music theory that are directly relevant to synthesis and sampling, as well as many other musical applications. While a comprehensive discussion of music theory is far outside the scope of this book, the core components of music theory are discussed here: rhythm, harmonic and non-harmonic sounds, musical intervals, and chord construction.

4.1.1 Rhythm

Rhythm is time divided in a musical context. The core units of time describing musical events are the bar (also called the meter in the US) and beat, defined by the time signature. A time signature is made up of two numbers written as a fraction – for example, 4/4.

The lower number of a time signature indicates the beat division. In 4/4, this is a quarter-beat (also known as a crotchet). The upper number shows how many of these beats there are in a bar. There are four quarter-notes in a bar of 4/4, counted as 1, 2, 3, 4.

4.1.1.1 Simple time

Simple time is a type of time signature where the standard division of each beat is split into even number divisions – that is, 2 × 8th notes or 4 × 16th notes. In contrast, compound time is split into three equal parts – generally groupings of 3 × 8th notes (explained in Section 4.1.1.2). Simple time signatures include 4/4, 3/4 (waltz time signature), 2/4 and 2/2.

For most practical uses, particularly in popular music, the typical lowest subdivision of a bar with a time signature of 4/4 can be made by splitting the bar into 1/16ths. Standard drumming terminology is to count these sixteenths as 1 e & a, 2 e & a, 3 e & a, 4 e & a.

The standard beat underpinning the vast majority of modern popular music (especially rock) is typically a variation of the kick drum on the 1 and 3, snare drum on the 2 and 4 (the backbeat), and closed hi-hats playing 1/8ths (1 &, 2 &, 3 &, 4 &).

Two other notable rudimentary rhythmic patterns in popular music are:

- 'four to the floor' – a kick drum on every beat and the hits described in Figure 4.1. Variations of four to the floor beats are common in electronic popular music. Figure 4.1 shows one bar of a four to the floor rhythm in the Ableton Live MIDI editor, using note lengths of 1/16ths; and
- 'breakbeat' (or 'break') – an embellished variation of the beat described in Figure 4.1, with many hits occurring away from the strong beats (playing notes or hits off the main beats is called syncopation). Breakbeats are very common in original RnB, funk and early hip hop.

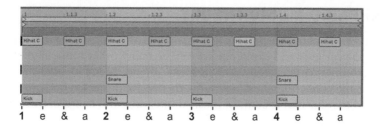

FIGURE 4.1

MIDI editor from Ableton Live showing a standard four-to-the-floor pattern including a backbeat and the musical count beneath the notes

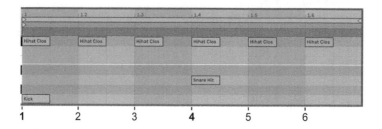

FIGURE 4.2

MIDI editor from Ableton Live showing a standard rock beat in a 6/8 pattern including the musical count beneath the notes

4.1.1.2 Compound time

Compound time signatures split the beat into three parts, which are counted as 1, 2, 3, so that in a bar or 6/8, there are only two beats, accented on the 1 and 4 – so with the accent shown in bold, the count is **1**, 2, 3, **4**, 5, 6. Figure 4.2 shows one bar of a typical 6/8 rhythm with a backbeat in the Ableton Live MIDI editor.

An example of a song with a 6/8 time signature is 'The House of the Rising Sun' by The Animals. Other examples of compound time are 3/8, 9/8 and 12/8. It is worth noting that 12/8 is often interchangeable with music played in 4/4 with triplets (triplets are three evenly spaced hits or notes typically played where they would generally be two).

4.1.1.3 Odd meter

Music with an odd meter combines both simple and compound time rhythmic patterns in a bar. 5/4 is an example of an odd meter time signature, where the rhythmic patterns can be arranged in either 3+2 (three beats followed by two beats, accented: **1**, 2, 3, **4**, 5), or 2+3, (accented: **1**, 2, **3**, 4, 5). The Jazz piece 'Take Five', by the Dave Brubeck Quartet, is an example of a piece of music in 5/4 where the feel of the rhythm is 3+2.

Other notable odd meter rhythms include variations of time signatures with 1/8ths in the lower number of the time signature – for example, 7/8; the rhythmic

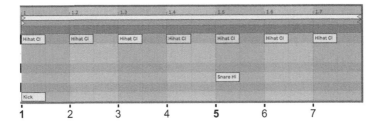

FIGURE 4.3

MIDI editor from Ableton Live showing a 7/8 rhythm with a 4+3 arrangement including the musical count beneath the notes

feel is typically arranged in 3+4 or 4+3 grooves. Figure 4.3 shows a 7/8 rhythm with a 4+3 rhythmic arrangement in the Ableton Live MIDI editor. The rhythmic arrangement is 4+3, because the snare hit lands on the 5 beat.

4.1.2 Harmonic and non-harmonic sounds

Every pitched note has a fundamental frequency, its loudest and typically lowest frequency component. (In contrast, percussive sounds, such as a snare drum or a cymbal hit, have a broad spectrum of frequency energy.) As mentioned in Section 1.1.1.2, any sound other than an ideal sine wave contains frequencies as well as the fundamental. These additional frequencies are called overtones or partials, which make up the sound's timbre or character. Overtones can be divided into three types: harmonic, non-harmonic, and noise.

4.1.2.1 Noise

In audio engineering, noise is typically an unwanted sound that accompanies an audio signal (see Section 1.2.1.2.1). In synthesis and instrument acoustics, however, noise is a random, non-musical sonic component that may give the sound an edge, a hiss, or even an underpinning bed. In synthesis, the type of noise that can be added to sounds is typically white noise, a random sound with equal intensities across the spectrum of human hearing. Because of the sensitivity of the human ear to high frequencies, white noise sounds like a high-frequency hiss, similar to the sound of the sea surf.

Other types of helpful noise in music production are pink and brown noise. Pink noise has less high-frequency content than white noise as it reduces 3dB for every octave, so it sounds richer than white noise and less hissy. Brown noise has even less high-frequency content as it reduces 6dB for every octave, giving it a deeper rumbling character.

4.1.2.2 The harmonic series

Harmonic sounds are sounds directly related to each other both musically and mathematically. In musical terms, harmonic sounds are consonant; they are harmonically stable, resonate well together, and sound musically comfortable. In contrast, non-harmonic sounds are dissonant – an unstable harmony that causes the sounds to clash. Dissonance is not necessarily bad, as effective use of dissonance can help to provide a range of musical and sonic effects ranging from intriguing or tense to complete horror.

The mathematical harmonic relationship is easier to describe objectively: harmonic sounds are those with frequencies that match the harmonic series. The harmonic series is a series of frequencies that are whole number multiples of the fundamental frequency. For example, if the fundamental frequency is 100 Hz, it will have harmonics of 200 Hz, 300 Hz, 400 Hz, 500 Hz, and so on.

The harmonic series is a fundamental musical and sonic concept that underpins a comprehensive understanding of how to create and manipulate textural, complex, and polyphonic sound. An awareness of the harmonic series is also beneficial to add further musical and acoustic context to the live sound and music production techniques discussed in Chapters 2 and 3, respectively.

If a composer wants a wide-open cinematic swell of sound, orchestrating a harmony based on the lower levels of the harmonic series will give a broad and consonant expanse of sound. At its core, the harmonic series is a simple concept that elegantly connects foundational elements of sound creation, music production, and music theory. This section breaks down the first seven harmonics of the harmonic series and the vital musical relationships within.

Using a specific example gives a chance to see the practical implications of the harmonic series. If a bass guitar plays the note A1 55 Hz (the A string), then the strongest vibration of that string will be 55 times a second (Hz). However, the string will vibrate at several harmonics of that fundamental. As well as 55 Hz, there will also be a series of additional notes typically vibrating with less energy. The amount of these additional frequencies and how they interact over time determine the character of the overall notes. For example, notes played on old flat-wound strings on a large-bodied bass guitar are likely to vibrate with less energy in the high harmonics and have a deeper tone. Notes played on new nickel wound strings on a smaller bass guitar are likely to vibrate with more energy in the high harmonics, so they are likely to have a brighter tone.

Unsurprisingly, these harmonics form the harmonic series. The first seven harmonics of A1 55 Hz are shown in Figure 4.4, showing the notes, frequencies, musical relationships, and a visual representation of each the string vibration.

First harmonic: The first harmonic, H1, is the fundamental – the note A1 – and is likely to be significantly louder than the other harmonics. In music theory, if this is the first note of the musical scale or key, it is called the tonic and establishes itself as a musical resolution. When the tonic is played after other notes, it sounds like the music has returned home.

Second harmonic: The second harmonic is another A, the note A2, with the same musical identity but an octave above the fundamental. The even-order harmonics (2,4,6, etc.) all have octave relationships with the lower harmonics, so they share a more direct musical relationship with the fundamental.

Third harmonic: The third harmonic is the first new musical identity, the note E3 (as mentioned, note numbering increases at C: A2, B2, C3, D3, E3, and so on). The third harmonic is always a perfect fifth plus an octave above the

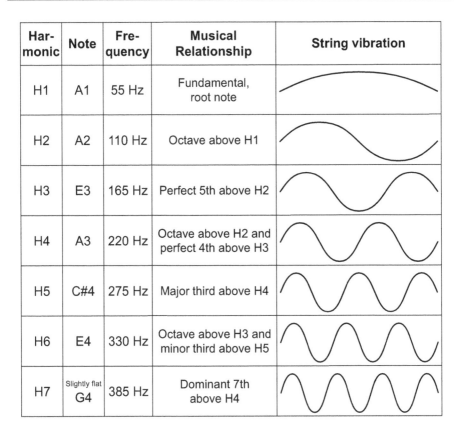

Har-monic	Note	Fre-quency	Musical Relationship	String vibration
H1	A1	55 Hz	Fundamental, root note	
H2	A2	110 Hz	Octave above H1	
H3	E3	165 Hz	Perfect 5th above H2	
H4	A3	220 Hz	Octave above H2 and perfect 4th above H3	
H5	C#4	275 Hz	Major third above H4	
H6	E4	330 Hz	Octave above H3 and minor third above H5	
H7	Slightly flat G4	385 Hz	Dominant 7th above H4	

FIGURE 4.4

The first seven harmonics of A1 55 Hz

fundamental. In Western music theory, this is known as the dominant degree of the scale because it functions as the next most significant scale degree after the tonic. It is no coincidence that the dominant forms a vital part of a note's timbre, as it is likely to resonate prominently.

Fourth harmonic: The fourth harmonic is another A, an octave above H2 – the note A3. The acoustic spacing of the harmonics is getting closer. In the first octave, A1-A2, there are no harmonics other than A1 and A2. The octave A2-A3 has one harmonic other than A2 and A3, the note E3. As the eighth harmonic is A4, then in A3-A4, there are three harmonics other than A3 and A4. (Incidentally, A4 is the pitch 440Hz, the calibration pitch for concert tuning, but it is not included in Figure 4.4 as it does not go to H8.) This acoustic spacing reflects the non-linear perception of frequency discussed in Section 1.1.1.2 and shown in Figure 1.3.

Fifth harmonic: The fifth harmonic is the second new musical identity, the note C#4 (C# being the black key on a piano keyboard immediately to the right of the C key). In an unfortunate quirk of terminology, the fifth harmonic is always a major third plus two octaves above the fundamental (notice that the third harmonic is a perfect fifth). These three musical notes – A, C#, and E – are also the notes of an A major chord. This relationship demonstrates why simple triads are so harmonious: the notes C# and E are likely already ringing out quietly within the note itself as overtones.

Sixth harmonic: The sixth harmonic is another E – specifically E4, the octave above the third harmonic.

Seventh harmonic: The seventh harmonic shows the difference between tuning by harmonics, called just intonation, and the standard method of tuning, called equal temperament. In equal temperament, a G4 is 392 Hz, but as we can see, the fourth harmonic is 385 Hz, a slightly flat standard G4. If an instrument were tuned to just intonation, it would need to be tuned to the musical key but sound particularly harmonious when played. For any other key, however, the notes would not sound as harmonious, which is why the standard system of equal temperament, while not as harmonious as just intonation, allows greater flexibility in performance. It is worth noting, however, that in most acoustic instruments, only the lower harmonics, 1–5, are likely to be loud enough to make a significant difference to the note's timbre.

4.1.2.3 Non-harmonic sounds

Any sound that is not harmonically related – that is, which does not have a frequency that is a whole number multiple of the fundamental – is considered non-harmonic. Compared to the harmonic sounds that resonate well together and sound musically pleasing, non-harmonic sounds clash musically and are likely to have a dissonant, harsh, or clanging quality. Non-harmonic sounds can be highly effective musically, as tension and dissonance are vital to many types of music. Control over the balance of consonance and dissonance of an instrument's timbre and the overall harmonic balance of a musical piece is hugely beneficial. Essentially, the further away from a harmonic relationship to a fundamental that a sound is, the more dissonant and harsh it will sound.

4.1.3 Musical intervals

Musical harmony is the combination of musical notes (rather than melody, where the notes are played in sequence). The distances between notes are called intervals, a critical characteristic that determines the type of musical harmony. If we have more awareness of the harmonic content of the notes, there is likely to be more awareness and effectiveness in the musical harmony created.

For example, a distorted electric guitar note has a much greater harmonic content than an undistorted or clean tone. The harmonic content of distortion is one of the main reasons that fifth chords – also called power chords – are so effective in rock music because a fifth chord is made up of the root note and the fifth note. For example, A^5 is made up of the notes A and E. A fifth chord lacks a musical third interval, a critical note in a triad, either a major or minor chord. For example, the chord A major is made up of A, C#, and E. See Section 4.1.4 for an explanation of chord construction. A large part of a chord's emotional characteristics is determined by the third, which also determines whether the chord is major or minor. Without the third, the chord sounds emotionally ambiguous or powerful. In turn, this makes space for the harmonics created by distortion, producing the characteristic 'big' rock sound (along with the dynamic range compression inherent in distortion discussed in Section 2.2.4).

Musical intervals are measured in semitones. One semitone is a single note increase on a piano keyboard – for example, the white note C to the black note C#. There are 12 semitones in every octave, and the semitone is broken down into cents (hundredths) for microtonal sounds or tuning purposes. There are typically three tuning levels for each oscillator: the octave, the semitone, and the cent (a hundredth of a semitone).

Musical intervals are discussed here and shown in Table 4.1 in the order of the number of semitones to help give musical context to sounds created using synthesisers and samplers (or effects that may change the pitch). The general musical characteristics of the 12 intervals that make up Western harmony are discussed in the comments of Table 4.1. These comments are considered

TABLE 4.1

The 12 musical intervals of western harmony

Semitones	Musical interval (using C major)	Harmonic comments	Common melodic identifier (the first two ascending notes of the tune)
0	Unison	Playing two with fundamental frequencies that are the same or very close is a standard simple method to make a fuller sound (although if the sounds are identical, then all that will be achieved is a volume increase of 3 decibels (dB)).	
1	Minor second (C-Dd)	A sharp dissonance useful for creating tension or unease in a sound and likely to beat quickly.	The *Jaws* theme
2	Major second (C-D)	A mild dissonance, less clashing than 1 semitone (a minor second).	'Frère Jacques'
3	Minor third (C-Eb)	A soft consonance typically regarded as melancholy. If a vocal harmony is made using intervals of three or four semitones (i.e., thirds), it is considered close harmony.	'Greensleeves'
4	Major third (C-E)	Another soft consonance, a close harmony that is typically regarded as happy compared to three semitones (minor third). Directly related to the fifth harmonic (discussed in Section 4.1.1.2).	'Kumbaya'

(continued)

▶ TABLE 4.1 (Continued)

Semitones	Musical interval (using C major)	Harmonic comments	Common melodic identifier (the first two ascending notes of the tune)
5	Perfect fourth (C-F)	Either consonant or dissonant depending on the musical context. Notably, five semitones is the musical inverse of a perfect fifth (i.e., C-F is 5 semitones, a perfect fourth, and F-C is 7 semitones, a perfect fifth).	'Amazing Grace'
6	Augmented fourth (C-F#) or diminished fifth (C-Gb)	An unresolved and ambiguous sounding interval, but typically dissonant, called the tritone (as it is three whole tones).	*The Simpsons* theme (augmented fourth)
7	Perfect fifth (C-G)	An open consonance directly related to the third harmonic (discussed in Section 4.1.1.2).	The *Star Wars* theme
8	Minor sixth (C-Ab)	A soft consonance but with wider harmonic spacing than three semitones (the minor third), so it sounds more open.	'Earth Song' (Michael Jackson)
9	Major sixth (C-A)	Another soft consonance but with wider harmonic spacing than four semitones (the major third), so it sounds more open.	'My Bonnie Lies Over the Ocean'
10	Minor seventh (C-Bb)	A mild dissonance but with a much wider harmonic spacing than two semitones (the major second), so it sounds more open.	The *Star Trek* theme
11	Major seventh (C-B)	A sharp dissonance but with a much wider harmonic spacing than one semitone (the minor second), so it sounds more open.	'Take On Me' (A-ha)
12	Octave	An open consonance and the second harmonic relative to the fundamental (discussed in Section 4.1.1.2). Twelve semitones (an octave) is always a frequency that is double the fundamental, and the most consonant of the intervals (other than unison).	'Somewhere Over the Rainbow'

generalised guidelines as they are discussed here in isolation from the musical context that is vital to establishing the role and character of the notes. The corresponding first two notes of very familiar melodies are also included to help identify the sound of the interval.

If consonance and dissonance are considered the extremes of complete harmony and a complete lack of harmony, respectively, then between the two are varying degrees of both consonance and dissonance. While this grey area is somewhat subjective, certain musical conventions use this resolution or tension. These are referred to in Table 4.1.

Beating: The most dissonant sounds are typically far enough apart in pitch to be distinguished as different notes but close enough to clash, so in the list of musical intervals, this will be one semitone, a minor second. When musical pitches are very close, the frequencies create an interference pattern called beating, which is perceived as a repetitive raising and dropping of volume – that is, tremolo. As the frequencies move closer, the beating rate slows until the pitches are the same when it stops. As the frequencies of the notes move apart, the rate speeds up until the beating is no longer discernible.

4.1.4 Chord construction

As introduced in the first part of Section 4.1.3, musical chords are a set of musical notes played simultaneously. The most common chord type in Western music is the triad, consisting of three notes: the root and the musical intervals of a third and a fifth from the root note.

For example, the chord C (also called C major; if there is no major or minor included in the symbol, then major is assumed) is made up of a C, an E (4 semitones, a major third), and a G (7 semitones, a perfect fifth).

Major and minor chords are by far the most common in Western music. As discussed in Section 4.1.2, the type of musical third decides whether the chord is major or minor. Because the chord C is made up of a major third interval (as well as a perfect fifth), this is a major chord. In contrast, the chord Cm (the 'm' or 'min' stands for minor) is made up of a C, an Eb (3 semitones, a minor third), and a G (7 semitones, a perfect fifth).

It is also helpful to appreciate the intervals between the notes themselves. For a major chord, such as the A chord shown on the left of Figure 4.5, the interval between the A to C# is four semitones – a major third; and the interval between the C# to E is 3 semitones – a minor third. The opposite is true of a minor chord, such as the A minor chord shown on the right of Figure 4.5: the interval between the A to C is three semitones – a major third. The interval between C to E is four semitones, a major third.

There are many other kinds of chords, variations on triads and extensions, for example, all with different harmonic functions.

4.2 MIDI

As introduced in Chapter 2, MIDI is a standard of control data. It is worth repeating that MIDI is not a representation of sound; it is the instruction to play sound (or to control some other system, such as lighting). Digital audio, however, is a representation of sound.

In a typical basic DAW setup, a plugin instrument will be triggered and controlled by MIDI data, creating digital audio as its output. The MIDI 1.0

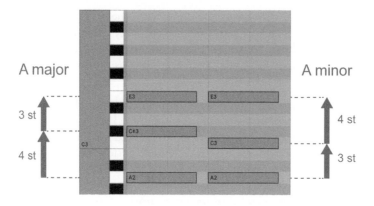

FIGURE 4.5

The intervals between the notes of A major and A minor shown on a MIDI piano roll

standard was released in 1983 and is the set of rules that define how MIDI messages work. Including modifications such as General MIDI (GM) (discussed in Section 4.2.1.1.2) and MIDI polyphonic expression (MPE) (discussed in Section 4.2.1.1.3). MIDI 1.0 is, at the time of writing, the most common system for controlling digital instruments and is a rare example of a robust but highly flexible, non-proprietary, but almost universally implemented standard. The weaknesses of MIDI 1.0 are mainly around expressive control and are discussed in Section 4.2.3.

As discussed in the introduction, musical expression is at the core of the vital exchange of energy in musical performance events. The term 'expressive' is used here to describe the fine expressive control inherent to acoustic instruments such as the violin or flute. As discussed in Section i.1.4, a violinist has a high degree of fine expressive control over each monophonic note's pitch and dynamics as they are played, compared to a pianist with more limited control but a much higher polyphony.

MIDI 1.0 messages are a series of instructions in three groups of eight-bit bytes, using seven of these bits to give a control range (providing the typical MIDI range 0-127, explained further in Section 4.2.1).

For example, a typical MIDI message would be 1001000 00111100 00110100. 1001000 is a 'note on' instruction in channel 1, which may have the patch of a lead synth loaded. 00111100 is the instruction to play a middle C (the MIDI pitch number 60). 00110100 is the instruction to play quite loudly (a velocity of 100). MIDI 1.0 messages are explained in further detail in Table 4.2.

4.2.1 MIDI 1.0 note messages

In MIDI 1.0, note messages comprise three bytes (each is an eight-bit digital word). The configuration of these bytes and several examples can be seen in Table 4.2. The first byte is the status byte and defines the use of the following data bytes.

The most common types of MIDI 1.0 messages for playing music are the note on, note off, pitch bend, and control change messages. The note on message starts the playing of a note, and the note-off message (or a note-on message with a velocity of 0) triggers the release section (see Section 2.4.1) and ends the note. MIDI 1.0 is a highly effective system of musical instruction, as only two three-byte messages are needed to start and stop a musical note. However, these note-on and note-off messages cannot change the note as it continuously plays.

Pitch bend and control change messages are continuous controllers (CCs) that change a note's specific characteristics over time. Pitch bend, as the name suggests, changes the pitch of the note, allowing vibrato and note slurs. Control changes allow a wide range of changes to a note within the instrument plugin, such as volume (CC 7) and panning (CC 10). General modulation (CC 1) depends on the instrument and patch but is typically either a vibrato or tremolo effect. Table 4.2 shows the configuration of these MIDI 1.0 messages.

The configuration of these messages shows and explains some of the basic rules of creating music using MIDI 1.0 messages. Each instrument sound or 'patch' is

---

TABLE 4.2

Table showing the configuration of MIDI 1.0 messages with examples

Status Byte	Data Byte	Data Byte	Description
1001nnnn	0kkkkkkk	0vvvvvvv	A status byte starting 1001 is a note on message, **nnnn** digits showing the channel number (1-16), **kkkkkkk** the note/key number (0-127), and **vvvvvvv** the velocity level (0-127)
10010000	00111100	01100100	A note on message for channel 1, note 60 (middle C, C4) velocity 100
1011nnnn	0ccccccc	0aaaaaaa	A status byte starting 1011 is a control change message, nnnn digits showing the channel number (1-**16**), **ccccccc** the controller number (0-127), and **aaaaaaa** the controller amount (0-127)
10110010	00000001	01000000	A control change message for channel 2, controller 1 (modulation), amount 64
1110nnnn	01111111	0mmmmmmm	A status byte starting 1110 is a pitch bend message, nnnn digits showing the channel number (1-16), **1111111** the least significant bit (fine pitch **adjustment), and mmmmmmm** the most significant bit (course pitch adjustment). 14 bits of data allows for a much higher resolution of control (0-16383)

controlled by a MIDI 1.0 channel, and as shown in Table 4.2, the last four bits of the status byte set the MIDI 1.0 channel, so there are 16 channels per MIDI 1.0 port.

As digital bits are binary, a base two number system (i.e., there are two numbers, either 0 or 1), for every position, the number increases by a magnitude of two (in contrast to the standard decimal system where the position increases by a magnitude of ten). A one-bit binary system has two numbers or levels: 0 and 1. In a two-bit system, there are four numbers/levels:

00
01
10
11

In a three-bit binary system, there are eight numbers/levels:

000	100
001	101
010	110
011	111

In a four-bit binary system, there are 16 numbers/levels:

0000	0100	1000	1100
0001	0101	1001	1101
0010	0110	1010	1110
0011	0111	1011	1111

This doubling of the numbers/levels continues and gives us 128 numbers/levels for a seven-bit system.

So, there are 16 MIDI 1.0 channels for every MIDI 1.0 port because four bits are allocated. Also, there are 128 notes and velocity levels (0-127) because seven bits are allocated. Each port is a physical input for hardware devices using the old-style MIDI 1.0 5-pin DIN inputs (described in Section 1.2.1.2.2). As Universal Series Bus (USB) is a much faster connection type, there can be many virtual ports depending on how the system is set up.

4.2.2 General MIDI

In 1991, the MIDI Association published GM, a standardised specification for MIDI 1.0 sounds. GM determines the names of the patches within a MIDI 1.0 channel and the percussion sounds on channel 10, which is reserved exclusively for percussion in GM. For example, in a GM drum map, on channel 10, note 24, or C1, is 'Bass Drum 1', and note 26, or D1, is 'Acoustic Snare.'

4.2.3 Restrictions of MIDI 1.0

Despite the stability, flexibility, and popularity of MIDI 1.0, it has two significant drawbacks: control change resolution and channel-wide CC. Strategies to work around these drawbacks are dealt with in the following two sections, but in the longer term, it is hoped that the creation of MIDI 2.0 (and hardware devices that make use of MIDI 2.0) will remove these drawbacks (see Section 4.2.5).

4.2.3.1 Control change resolution

As seen in Table 4.2, a control change message is allocated a seven-bit word length, which gives only 128 steps of resolution. While this is not usually a problem, for large changes to a sound's parameter, this is not likely to provide a smooth transition over a long period (e.g., an exaggerated sweep of the low pass filter cutoff). The transition may sound stepped, like a zip being closed.

Pitch bend is the standard MIDI method to change the frequency of a note by a small musical amount. Because the human ear is sensitive to pitch changes, pitch bend is given its own specific continuous controller type (as shown in Table 4.2). Accordingly, pitch bend does not need the first data byte to specify controller type and uses both data bytes to allow two seven-bit words for the resolution. The first data byte is the most significant byte (MSB), which is the byte with the highest value, so its position, when written down, is to the left. The second data byte is the least significant byte (LSB), which is the byte with the lowest value, so its position, when written down, is to the right. Between every step of the MSB, there are 128 steps of the LSB, allowing a total of 16,384 possible steps.

There are two main workarounds to the low resolution of control change:

* CC pairs
* non-registered parameter numbers (NPRNs).

Both options allow 14-bit MIDI 1.0 control messages but manage this differently.

CC pairs: CC pairs combine two CC messages, typically CC 0-31 for the MSB and another CC that is 32 higher (CC 32-63) for the LSB. For example, the Moog Voyager synthesiser uses the CCs 19 and 51 for its filter cutoff. While the MIDI 1.0 specification CCs 32-63 are set aside to be LSBs for the CCs 0-31, these pairings are mostly left undefined and are not widely or easily implemented by DAWs, hardware instruments or plugins.

NPRNs: NPRNs are CC messages set aside as non-registered by the MIDI 1.0 standard, allowing instrument-specific modes of operation. For example, the CC pair CC 98 and CC 99 sets the parameter, and the CC pair CC 6 and CC 38 sets the value. The first CC is the LSB, and the second is the MSB, allowing 14 bits for both parameter and value.

4.2.3.2 Channel-wide CC

Any control change or pitch bend message will apply across the channel (as seen in the format of the CC messages shown in Table 4.2), affecting any notes being played on that channel.

Two different examples are helpful to show the typical results of this restriction.

The first example involves the recent development of more expressive controllers – for instance, the Haaken Audio Continuum Fingerboard, the Roli Seaboard, and the Linnstrument, commercially released in 1999, 2013 and 2014, respectively and discussed further in Section 4.3.2.4.

A synthesiser played by a Roli Seaboard is an example that illustrates the issue in this context. The Roli Seaboard is essentially a piano keyboard layout with keys made of a spongy rubber that allows pressure, and vertical and horizontal control (as well as how each note is hit and released), depending on finger movement. While it is possible to map each of these movements to different parameters of a synthesiser, a typical mapping would be a horizontal movement for pitch (pitch bend), pressure for volume (CC 7), and vertical movement for modulation (CC 1). Using standard MIDI 1.0 messages on one channel means that the most recent CC message will be applied to all notes, ignoring any previous CC messages of the same number used for other notes played before that are still sustaining. This restriction is not a problem for a monophonic synthesiser (synthesisers that can only play one note at a time, often abbreviated to mono synths and discussed in Section 2.3.5a). However, this is a significant limitation for polyphonic synthesisers (poly synths).

The second example involves the programming of MIDI 1.0 into DAWs. In this case, a string ensemble played by a sampler, such as Kontakt, is an appropriate example to illustrate the issue in this context. A sample library in Kontakt is likely to control dynamics using CC1 (despite being labelled as modulation) and volume using CC7 (channel volume) or CC11 (expression, typically a proportion of channel volume). These are generalisations, however, and different libraries may use other methods. The real drawback of this scenario is if that string ensemble is playing a harmony (e.g., the notes C2 and G3) on the same channel and requires a dynamic swell or crescendo, any message to change dynamics on CC1 will change both notes at precisely the same time, which is likely to sound artificial.

The traditional solution to this problem is to put the instrument into MIDI mode 4, Omni Off/Mono, otherwise called mono mode (rather than the standard MIDI mode 1, Omni On/Poly, which responds to MIDI 1.0 data regardless of channel, and is polyphonic per channel). MIDI mono mode assigns a different MIDI 1.0 channel to each note played and only allows one note per channel, so that MIDI 1.0 CC messages are per note. MIDI mono mode is effective but provides a maximum of 16 voices.

Another unique solution to the problem of editing polyphonic expression is version three of Steinberg's virtual studio (VST) instruments plugin interface, released in 2008. VST3 allows high-resolution and polyphonic expression, but is only fully supported within the Steinberg DAW's Cubase and Nuendo. While this makes for powerful editing functionality within the DAW, the features of VST3 plugins that extend the MIDI 1.0 standard are, unfortunately, incompatible with expressive controllers. Instead, these controllers make use of MPE.

In 2018, the MPE enhancement was added to the MIDI 1.0 standard, which gives a more standardised and sophisticated solution to MIDI 1.0 mode 4/ mono mode.

4.2.4 MPE

As the name suggests, MPE is a solution to limiting channel wide/monophonic expression. MPE assigns a different channel to each note played, much like MIDI 1.0 mode 4/mono mode; however, when it reaches 16 notes, more notes are allowed as the channels can be polyphonic. Notably, the 16-channel limit cannot be overcome, so CC messages are shared from this point.

In practice, MPE means that expressive instruments and MIDI editing are becoming far more accessible and standardised. An example of MPE editing in Ableton Live 11 is shown in the expression tab of the MIDI editor piano roll in Figure 4.6. In Figure 4.6, an A major chord has the C#3 pitch bent down to a C3, making the chord an A minor. The pitch bend is clearly shown in the four connected red circles because the C# note is selected, but other expression edits can be seen in the grey plots of the slide and pressure lanes.

The slide lane shows the alteration of A2, which has the effect of raising the cutoff of the filter to make the tone of A2 brighter and harsher. The pressure lane shows the alteration of E3, which changes the wavetable positions and fold settings, giving additional timbral distortion. The mappings that create these tonal changes are shown in Figure 4.7 (in the MPE Modulation Matrix of the Wavetable Plugin by Ableton).

These expressions would not be possible in a single track with standard MIDI 1.0 (i.e., without MPE).

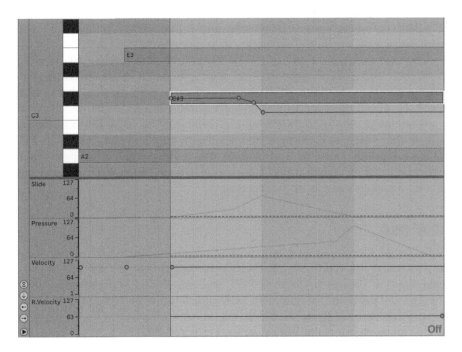

FIGURE 4.6

Expression tab of the MIDI editor in Ableton Live with note C#3 selected

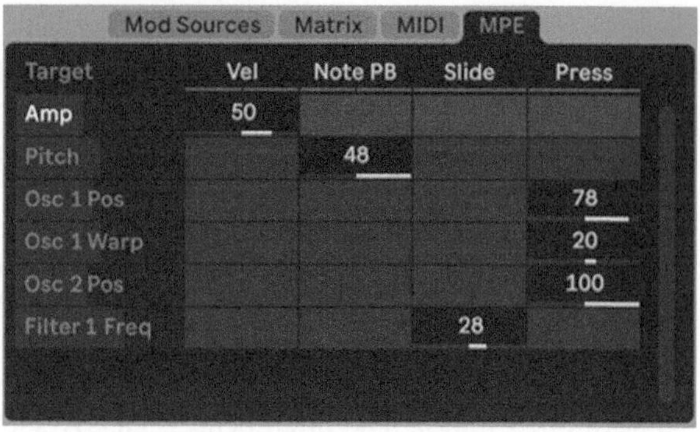

FIGURE 4.7
MPE Modulation Matrix of the Wavetable Plugin by Ableton

The finger movements typically associated with these expressions are discussed in Section 4.3.2.5.1.

4.2.5 MIDI 2.0

In January 2019, the MIDI Manufacturers Association (MME), a non-profit trade association for companies that develop products that use MIDI technology, announced the prototyping of MIDI 2.0. The press release stated[1]: "The MIDI 2.0 initiative updates MIDI with auto-configuration, new DAW/web integrations, extended resolution, increased expressiveness, and tighter timing – all while maintaining a high priority on backward compatibility."

The 'auto-configuration' and 'new DAW/web integrations' of MIDI 2.0 are likely to bring about some exciting developments. It is somewhat inevitable that the shiny new features of MIDI 2.0 are more prominent than the problems solved. However, these are all welcome developments.

While bold new announcements are commonplace in audio technology, it is likely that MIDI 2.0 will be a significant new development. Although MIDI 1.0 has admirably stood the test of time, it is understandably hampered by the restrictions of 1983 when it was released. It is hoped that MIDI 2.0 will be as successful and effective as its predecessor.

4.2.6 OSC

Open Sound Control (OSC) is a notable alternative to MIDI, developed by the University of California, Berkeley's Centre for New Music and Audio Technology. OSC was released in 1997 and uses a URL-style message system. OSC is more open and flexible than MIDI but has nowhere near the amount of industry takeup or support. As it is based on network protocols, OSC does lend itself to more experimental WiFi systems. However, WiFi is far less likely to offer the reliability needed for live performance.

4.3 HARDWARE

The physical equipment used to perform modern electronic music ranges from computers and all the associated peripherals (e.g., controllers and audio interfaces) to standalone instruments, such as electronic piano keyboards (referred to from this point on as just 'keyboards'). As well as computers and standalone instruments, there is a wide range of modular equipment for sound manipulation and creation that can be combined, such as Eurorack synthesiser modules and effects units.

Section 4.3.1 discusses the use of computers and leads onto Section 4.4, which discusses the accompanying software.

Section 4.3.2 discusses controllers: hardware devices designed to trigger and control musical or system events. Importantly, these are peripheral hardware devices and, on their own, cannot create music.

Section 4.3.3 discusses standalone electronic instruments, from keyboards to less familiar instruments such as the Continuum Fingerboard. Most of the equipment discussed in this section is based on digital technology due to its comparable flexibility, reliability, and low cost for a wide range of audio manipulation tasks. Purely analogue electronic instruments are typically vintage-style synthesisers. However, many hybrid instruments use a combination of analogue and digital technology, typically making use of analogue elements to give a degree of warmth and non-linearity – for example, in preamp stages.

Section 4.3.4 discusses effects units with a focus on live performance. In this context, there is a more even split between analogue and digital technology, mainly due to the smaller size of the effects units.

The various hardware devices that make up electronic music performance events are outlined in this section. In designing a setup of equipment for live performance beyond the required feature set, several main factors are worth considering. As for most live performance events, the most critical factor in setup design is likely to be reliability, while other factors are likely to include a combination of cost, feasibility, and individual preference.

Many different devices are available for electronic music performance, with much overlap in their sound manipulation techniques. Two distinctly different approaches to devices and their setup can be summarised as:

- a standalone hardware device or a combination of standalone hardware devices, and
- a computer with various peripheral hardware devices (e.g., including audio interfaces and MIDI controllers).

Standalone hardware device setups: A setup based on a single standalone device, such as a general-purpose keyboard, is likely to be the most reliable and convenient arrangement; but it is also the most restrictive, as the sounds and workflow are limited to just that hardware device and its sole operation.

A setup with multiple standalone hardware devices is likely to be more robust than just a single computer because they operate more independently and are easier to replace or remove if they fail. Also, multiple standalone hardware devices typically offer a more bespoke and tone-purist approach.

However, a setup with multiple standalone hardware devices is a high-maintenance and high-cost approach, which does not lend itself to more innovative techniques such as various control methods or a high level of integration. Consequently, this makes the setup more robust because each hardware device is likely to be most effective when used discretely.

Computer-based setups: In contrast to a setup based on one or more standalone hardware devices, a computer-based setup offers a vast array of performance and accompaniment options. However, as there is more reliance on a single central computer, there is more risk to the performance should the computer fail. As well as combining a computer setup with other standalone hardware devices to spread both the workload and the risk, methods of managing the risk of computer failure are variations of systems of multiple redundancies. For example, this could be running multiple systems in parallel with a panic crossfader and having a simple backup system such as a backing track. For these methods to work effectively, they would need to be synchronised, so a fixed backing track is likely held in reserve as a worst-case backup.

4.3.1 Computers

Any standalone digital hardware device (e.g., a stage piano) can be considered a specialised computer, in that it manipulates digital audio using microprocessors. What is meant here by the term 'computer' is a general-purpose hardware device that runs an enormous range of software programs, from web browsers to spreadsheets to DAWs. As such, computers in their various forms – from desktops to smartphones and tablets – have become embedded in the modern Western lifestyle.

As mentioned in Section 4.3, setups based on multiple standalone hardware devices tend to be more robust than systems based on computers – partly because systems based on various standalone hardware devices tend to restrict their operation to more established paradigms.

However, as computers become more powerful, the processing headroom increases, and if managed carefully, a computer can provide a feasible alternative to a large stack of standalone hardware devices. This alternative then opens the door to a range of frameworks for innovation, from more conventional software instruments to programming environments allowing the user to create control, manipulation, and hardware device interconnection.

To specify and maintain a machine optimised for the audio, it is worth being aware of a computer's key elements. The following three sections discuss a computer's processor, memory, and audio interface. (Graphics cards are not discussed here, as they are not particularly relevant for audio.)

4.3.1.1 Processor

A computer is a hardware device that performs mathematical operations on sets of numbers – that is, data representing various things, such as text, images, video, or audio. The core of a computer is the processor, the element that performs these calculations. A processor's performance is measured in its clock speed and type. For example, a processor may have a 3.2 gigahertz clock speed and be a 16-core intel i9 or an ARM M2. When specifying a processor, it is typically best to compare the various options and to opt for the highest specification possible, but without using any feature that may compromise its reliability, such as overclocking or turbo boost. It is worth noting that many Apple Macintosh computers use ARM chips that are integrated systems on a chip so that the processor also includes the random-access memory (RAM) and graphics processing.

4.3.1.2 Memory

To store, recall, and process data, a computer uses two types of memory: temporary and permanent. Temporary memory is referred to as RAM while permanent memory is typically stored on hard drives, specifically solid-state drives. RAM is a temporary memory that the computer uses to hold the data needed for the processes in use at that time and is typically made up of several integrated circuits (or 'chips'). A common analogy for RAM size is a physical desktop used for office work: the bigger the desktop, the more different files can be used simultaneously. Similarly, the more RAM, the more programs (including instruments or effect plugins) can be running at the time. A typical RAM amount for an audio computer at the time of publishing is 16 gigabytes (GBy) or more, with the RAM amount increasing in leaps of eight or 16.

In the analogy of the physical office, the permanent memory is the filing cabinets: the storage for files that are not in use. Recorded audio is likely to be saved directly to the hard drive, so fast data transfer is essential. (Samples may also be recalled from permanent memory if there is insufficient RAM.) It is also common practice to have multiple drives so that audio can be written to and read from a drive different from the system drive (the drive running the operating system).

Typical specifications for solid-state drives at the time of publishing are 500 GBy to 1 terabyte (size) and 500 megabits per second (speed).

4.3.1.3 Audio interface

General computer peripherals include the necessary QWERTY keyboard and mouse/trackpad. A standard component of most computers is an audio interface; however, inbuilt audio interfaces are not likely to be reliable, high quality, or fast enough for live performance events and music production.

A typical inbuilt audio interface is likely limited to CD audio standard, 16 bit 44.1 kilohertz audio or resolutions below (see Section 1.1.3.3). Also, the standard audio output connector is the 3.5-millimetre headphone jack, which – while fine for consumer headphones – is not likely to stand up to the rigours of live performance events for very long.

Audio interfaces have two main components: the microphone preamps and the converters (analogue-to-digital and digital-to-analogue, discussed in Section 1.1.3).

The microphone preamps amplify the small microphone signals, bringing them to line strength. Other input options often include line (for powered signals already at line strength) and instrument (for instruments with direct output, such as electric guitar and bass). There may also be a Hi Z option to allow for the high impedance of guitar pickups.

After the input signal has been set to an appropriate level using the gain setting, the converters change the analogue signal into a digital signal using the sample rate and bit depth typically set by the software – for example, the DAW.

Most audio interfaces fall under the categories of connection type. At the time of publication, the most common connections are USB and Thunderbolt, each with version numbers, with the highest version in 2022 being USB4 and Thunderbolt 4. A table of USB transfer speeds is shown in Table 1.3, and the connector types are shown in Figure 1.8.

USB 2 is likely the most cost-effective connection for interfaces between two and eight input and output channels. In contrast, USB 3, 4, and Thunderbolt audio interfaces can offer much higher track count with lower latency, but are likely to be more expensive.

Larger-scale setups may use separate preamps and converters, especially if a mixing desk is part of the setup because microphone preamps are typically an integral part of the mixing desk.

4.3.1.4 Desktop and laptop computers

There are two types of computers based on the degree of mobility required of the machine: desktop computers and laptop computers. Desktop computers are hardware devices that require little mobility, so they do not typically have a battery and are likely to have several separate units – particularly the keyboard and mouse – connected as peripherals. In contrast, laptop computers are designed to be portable, with a battery that can run for at least several hours and are likely to have the whole computer, monitor, keyboard, and trackpad built into one hinged unit.

Laptop computers are often associated with more experimental electronic performance, with art music ensembles often referred to as 'laptop orchestras'. A laptop computer is typically more robust than a desktop computer and can be moved about easily. However, the trade-off for the integrated and compact design is often a lower specification and difficulty in cooling when the processor comes into heavy use.

4.3.2 Controllers

As computers become increasingly popular and accessible, so does the associated peripheral equipment. In music, peripheral devices for playing, triggering, and controlling music are generally called controllers.

MIDI is the most common connection standard, over either USB or Bluetooth, for wireless hardware devices. A notable exception to MIDI connections for music is OSC (discussed in Section 4.2.6). Because OSC is a standard based on internet network protocols, it is well suited to touchscreen hardware devices such as tablets and smartphones. Wireless connections such as OSC over WiFi or MIDI over Bluetooth offer an opportunity to explore more freedom of movement; however, it is worth balancing this against the risk of low batteries and connection loss.

The standard types of MIDI controllers – keyboards and pad controllers – are discussed in the following sections, followed by a discussion of touchscreen controllers and QWERTY keyboards.

One particularly notable area of modern controller development is expressive controllers, which are explored in Section 4.3.2.5 (expressive synthesisers are also discussed in Section 4.3.3.1.1). As mentioned in Section 4.2, the term 'expressive' is used here to describe the fine expressive control inherent to some acoustic instruments (e.g., the violin or flute).

Keyboards and pad-based controllers have a standard set of control conventions that are explored in Sections 4.3.2.1 and 4.3.2.2. The further away from these controller conventions, the more varied the possibilities for mapping becomes, especially with the wealth of parameters available for mapping. On the one hand, this freedom allows nearly unlimited exploration of control and performance methods, so a better question may be what *not* to map. On the other hand, this lack of performance convention is likely to distance the audience from their appreciation of the manipulation techniques used.

One suggested solution to this problem is to communicate to the audience the systems in use so as not to hinder performance by becoming a technology demonstrator. Another solution is the shared use of modern controllers so that audiences understand these musical-cultural signifieds. Musical-cultural signifieds are concepts and meanings gained from the musical and visual indicators in performances, and are explored further in Section 5.1.3.5. A critical mass of musical-cultural signifieds may be achieved as controllers become more widely used; however, it is likely to stay an issue for the more innovative and less conventional controllers. (Further concepts of mapping and performance semantics are explored further with a more theoretical framework in my forthcoming academic book on liveness in modern music performance.)

4.3.2.1 MIDI controller keyboards

The most popular musical controller is the MIDI controller keyboard (as are the piano keyboard instruments mentioned in Section 4.3.3.1). Typically, MIDI controller keyboards are made up of either 25, 49, 61, or 88 'synth action' piano keys that sit to the right of a modulation wheel and a pitch bend wheel. 'Synth action' keys are a simple and cost-effective mechanism consisting of a key, a hinge, a spring, and two buttons. The first button is activated at the beginning of the key press and the second at the end. The time between the pressing of both buttons gives the note speed and explains the MIDI term 'velocity', which is equivalent to how hard the key is pressed.

The keys of the MIDI controller keyboard are arranged in the chromatic notes format of the piano. In most MIDI controller keyboards, the keys simply start and stop the notes or hits on an instrument patch. The modulation and pitch bend wheel give a limited degree of expression, with the modulation wheel defaulting to MIDI CC1 and the pitch bend wheel set aside for changing pitch. (See Section 4.2.3.1 for more details of MIDI CC messages.) For MIDI controller keyboards with less than 88 keys, there is often an octave offset function that offsets the pitch of the keys to higher or lower ranges.

The MIDI 1.0 message note on/note off format was designed principally around the piano keyboard, which gives some insight into its main strengths and weaknesses. Its strengths are its robust, flexible, and standardised nature (discussed in Section 4.2), and its weakness are the expressive limitations (discussed in Section 4.2.3).

As we would expect, MIDI controller keyboards are well suited to playing piano and organ tones, particularly when used with pedals. The popularity of MIDI controller keyboards is also due to them being an adequate, if not ideal, all-round control mechanism for other pitched instruments. The main drawback of keyboards as controllers, as with MIDI 1.0, is the lack of expressive control, particularly for polyphonic parts.

This drawback is particularly apparent in patches that recreate the tones of pitched instruments attempting to generate sounds that typically have a direct connection between musician and instrument, such as the violin and flute.

The drawback of a lack of expressive control is also apparent in pitched electronic instrument patches that encourage a high degree of sound manipulation – for example, synthesisers and samplers with deep manipulation and modulation capabilities. It is these capabilities that are so ideally suited to being mapped to performance actions for enhanced performer expression. Expressive control is explored further in Section 4.3.2.1.2.

It is worth noting that an early solution to lack of expression in MIDI controller keyboards was aftertouch – the continuously altering pressure applied to the key after the initial key press. Expressive controllers have mostly replaced polyphonic aftertouch in MIDI controller keyboards, but polyphonic aftertouch is quite common in pad-based controllers (discussed in Section 4.3.2.2), such as the Novation Launchpad X and Launchpad Pro.

4.3.2.1.1 Expressive acoustic (and electric) pitched instruments and plugins
A basic DAW setup can bring about the common situation where a violin patch is played using a basic MIDI 1.0 keyboard (as described in the previous section). While monophonic expression is possible using the standard pitch bend and modulation wheel, this is not likely to sound like a real acoustic violin. This section and the following section discuss the workarounds available for introducing more expression in performance in standard MIDI 1.0 setups. Setups with expressive controllers are explored in Section 4.3.2.5.

A typical solution to the problem of lack of expression using a basic MIDI 1.0 keyboard is the use of articulation sample sets. Articulations are different ways of playing the same notes on string, woodwind, or brass instruments. Each articulation is likely to be a set of samples of playing notes that typically fall into short articulations (e.g., the short duration and fast attack of staccato) and long articulations (typically variations on sustained notes). These articulations are likely to be controlled by MIDI notes outside the range of the instrument. (Some DAWs also allow articulations to be set within the MIDI editor, allowing polyphonic articulations.) Each of these articulations is likely to sound the same if repeated regularly; however, the modulation wheel (CC1 messages) can also give additional musical alteration.

This approach can be handy for monophonic instruments, but for polyphonic instruments played using physical techniques such as strumming or fingerpicking (e.g., the guitar), this adds another layer of complexity. Like articulations for bowed strings, several virtual guitarist plugins are available that allow the selection of strum patterns using MIDI notes outside the range of the notes. Then when the notes are played on the MIDI keyboard, the notes are strummed in the pattern selected.

While these plugins can never replicate the unique nature of an acoustic instrument played live, they offer a very efficient and convenient alternative that affords several possibilities for generative, reactive, or interactive accompaniment in live performance.

4.3.2.1.2 Expressive electronic pitched instruments and plugins

The fact that the piano-style keyboard is well established as the conventional control mechanism for synthesisers has long been a source of frustration for many interested in exploring the sound manipulation potential of synthesisers and samplers. The standard piano keyboard is not well suited to control the array of expressive sound manipulation parameters, such as filter cutoff, low frequency oscillator rate, wavetable position, and sample loop durations.

The typical solution for expanding the MIDI controller keyboard has been to add dials, faders, and fader strips so that parameters can be mapped and controlled as needed. While this may be effective for simple parts and single parameter changes at a time, this is quite a crude method for performance.

The modern solution to this problem is expressive controllers, discussed in Section 4.3.2.5.

4.3.2.2 Pad-based controllers

While not as commonplace as the MIDI keyboard controller, pad controllers are a particularly effective means of manually controlling percussion patches and pitched instruments where an alternative (but also widespread) format to the chromatic keyboard is required.

Pad controllers typically consist of a grid of either square or rectangular pads with other elements such as buttons, dials, faders, and a screen, depending on the device's functionality. The pad grid is typically used to play notes or hits in an instrument patch or clip launching. Other elements, such as faders and dials, are mappable to most DAW parameters. Specific use of the QuNeo pad controller is discussed in Section 5.4.3.2.

As well as the general flexibility of pad controllers, there is a specific advantage to not being tied to the chromatic format of the MIDI controller keyboard: any arrangement of notes can be mapped to the grid of pads. A usual chromatic pitch arrangement is to tune the pads like a guitar so that each pad increases by a semitone from left to right, and the rows increase by a musical fourth (an interval of five semitones, see Section 4.1.2) from the bottom up. For situations where a limited set of notes would be desirable for diatonic (seven note) or pentatonic (five note) scales, for example, the pads can be set up so that any unwanted notes are left off the pads. While limiting the notes on the pads reduces the scope for musical expression, it makes it easier to perform in-key. When microtonal scales (scales that include notes in between the Western standard equal temperament 12-note scales) are required, these can also be mapped as required.

The more fully featured pad controllers are also likely to offer polyphonic aftertouch so that the pad's pressure can be mapped to a parameter (e.g., dynamics or filter cutoff), allowing more expressive performance than a synth-action keyboard, for example.

Particularly notable pad controllers include the 4×4 pads of the MPC series of controllers by Akai, and the 8×8 pads of Push by Ableton and Launchpad by Novation.

4.3.2.3 Touchscreen controllers

Before the popularity of touchscreen hardware devices such as tablets and smartphones, the Lemur touchscreen controller, made by Jazzmutant, was a popular device. The Lemur offered a wide range of control templates and incorporated innovative physics engines allowing modelled physical behaviours (e.g., fader friction and sprung control handles). Notably, the original Lemur hardware used OSC rather than MIDI, which remained the standard connection for touchscreen hardware devices for some time.

As tablet and smartphone technology became widespread, any significant commercial demand for separate touchscreen hardware disappeared. Several iOS and Android apps allow OSC and MIDI over WiFi and wired (USB) MIDI control. These include the well-established TouchOSC, made by Hexler, which features a wide range of control templates and is accompanied by TouchOSC Bridge, a program that converts wireless OSC to MIDI.

Many more simplistic controller apps give more limited keyboard, sliders, dials, and pad options, such as 'MIDI controller' by Dominik Seemayr, or apps that allow access to other data available to mobile hardware devices (e.g., accelerometer data).

Aside from touchscreens as controllers, many standalone 'instrument' apps are also available for tablets, most notably Borderland Granular (iOS only). Borderland Granular makes good use of the touchscreen format for sound creation and can run in conjunction with Ableton Link (to synchronise with an Ableton Live set); however, there is still the problem of managing the audio from a mobile device.

The flexibility and accessibility of touchscreens make them incredibly powerful and easy to master. However, the opportunity to interact with a physical key, pad, or fader is significantly absent.

4.3.2.4 QWERTY keyboard

An honourable mention is worth giving to the ubiquitous QWERTY typing keyboard. The QWERTY keyboard was initially designed to slow down typing speeds to avoid jamming the typewriters' mechanism, making it a frustrating exemplar of a non-ideal controller. Once established, this QWERTY keyboard for typing standard has been extremely resistant to alternatives, much like the piano keyboard for playing synthesisers. The phenomenon of sticking firmly to familiar options is a known trend. It is known as 'path dependence in economics or' immunity to change' in psychology.

It is common for DAWs to offer a means to use the keys on a QWERTY keyboard to create MIDI notes at a specified velocity, typically for the situation where no MIDI controller keyboard is attached. Also, some DAWs, especially Ableton Live, allow key mapping (as well as the standard MIDI mapping). Key mapping makes the QWERTY keyboard particularly useful in a perform-ance environment for varied triggering purposes where pressure sensitivity is not required. For example, the flow machine performance setup of Tim Exile[2] includes a QWERTY keyboard, used to trigger glitch-type effects.

4.3.2.5 Expressive controllers

This section covers a selection of controllers offering solutions to the expressive limits of the standard MIDI controller keyboards (and, to a lesser extent, pad controllers). This list is not fully comprehensive, but the most notable and up to date (at the time of publishing).

4.3.2.5.1 Expressive MIDI controller keyboards

Expressive MIDI controller keyboards are a type of augmented keyboard con-troller based around the keyboard's design but using materials and mechanisms that allow significantly more continuous control and expression than equipment with standard velocity and aftertouch available on each key or pad.

Each of the expressive MIDI controller keyboards mentioned here (as well as the expressive synthesisers discussed in Section 4.3.3.1) uses different methods of creating touch data from a piano keyboard-shaped device. The control categories can be summarised as the five-finger movements in the table shown in Figure 4.8. An example of editing these per-note expressions is shown in Figure 4.6.

These additions to a standard MIDI controller keyboard aim to maintain the general playing techniques associated with the piano keyboard while offering additional expression.

Roli Seaboard: The Seaboard, made by Roli, is possibly the most well-known of the expressive MIDI controller keyboards and is also the least like a standard MIDI controller keyboard. It is made of a squashy dark-grey rubber material, with rounded raised areas in the layout of the piano keyboard. Above and below the key shapes, the rubber is a flat surface, allowing the finger to slide hori-zontally between various pitches, giving a smooth horizontal (typically pitch) movement up and down the key range.

Roli was instrumental in developing MPE (discussed in Section 4.2.4). While the Seaboard is effective for controlling third-party MPE instruments, it is par-ticularly well suited to the Roli synthesisers developed to be played with the Seaboard. The names Roli use for the five categories of control (referred to as 'dimensions of touch' by Roli) are Strike, Press, Glide, Slide, and Lift, respectively.

Name	Finger movement	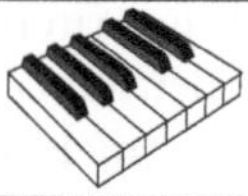	Typical mapping
Pitch bend (per note)	Horizontal movement		Pitch change
Slide	Horizontal movement		Timbral change (i.e. the frequency cutoff of a low pass filter)
Pressure	Continuous downward and upward movement		Dynamic or timbral change
Velocity	Speed of initial down-ward movement (i.e. the attack)		Initial dynamic
Release	Speed of upward finger		Release time or release sample

FIGURE 4.8

Expressive instrument finger movements, graphical representations, and typical mappings

Keith McMillen K-board pro: The K-Board Pro, made by Keith McMillen Instruments, is more recognisable as a keyboard variation than the Roli Seaboard and, at first glance, could be mistaken for a standard MIDI keyboard controller. The 'keys', however, are firm rubber pads (rather than the hinged keys of a MIDI controller keyboard), and allow the five control categories discussed previously in this section.

Two other expressive keyboard-based synthesisers – the Continuum Fingerboard and the Osmose – are discussed in Section 4.3.3.1.1. As they are not exclusively controllers, they are not included in this section.

4.3.2.5.2 Surface controllers

Other controllers that are not MIDI controller keyboards can be described as rectangular surfaces that are controlled manually in various ways.

QuNeo: Most modern pad-based controllers discussed in Section 4.3.2.2 transmit pressure (called aftertouch for keyboard controllers), data, and the standard note on messages. A notable pad-based controller with additional dimensions of expression is the QuNeo, made by Keith McMillen Instruments. Each pad on the QuNeo, as well as transmitting note on and pressure data, also transmits XY positional information, which makes it particularly useful for customising expression data using either MIDI 1.0 messages or MPE.

Linnstrument: The Linnstrument is a grid of small pads (either 8×16 or 8×25) that can light up and be considered an extended and expressive pad controller, as each pad has the five categories of control mentioned in Section 4.3.2.4.1. This wide grid arrangement is particularly well suited to tuning the rows to musical fourths (as mentioned in 4.3.2.2).

Overlay surfaces: Other surface controllers include Sensel's Morph and the Joué Board, created by Joué – both rectangular hardware devices with a flat surface designed to be used with a range of rubber overlays. These different overlays include a keyboard, pad, drum pad slider, dial, and knob overlays.

4.3.2.5.2 Game controllers

The gamepad is the most common gaming controller/input hardware device. It comprises a single unit held by both hands and operated with fingers and thumbs. While gamepads can be used to control instruments, they are not particularly well suited as they are designed to be used with a high degree of screen feedback. Two alternative game controllers developed in the early 2000s do, however, lend themselves particularly well to more experimental forms of music performance, as they make use of gestural and positional data. Respectively, these controllers are the Wii remote controller, made by Nintendo (generally known as the Wiimote), and the Gametrak controller, made by In2Games. These controllers are explored here, as are other gestural control systems such as the Leap Motion controller and the Roland D-beam.

Wiimote: The Wii and the accompanying Wiimote were released in 2006 and were a spectacular success, with over 100 million consoles sold. Rather than focusing on touch controls as gamepads do, the Wiimote uses accelerometers to detect movement in three dimensions, an infrared camera to act as a pointing hardware device, and several extra buttons for additional control. The Wiimote connects to the Wii by standard Bluetooth, which means that many other systems can use the Wiimote data. The combination of this physical control data, the flexibility of use, and the low cost of the Wiimote made it very popular to make music by controlling audio programming environments (e.g., Max and Reaktor, discussed in Section 4.4.5). The application OSCulator, by Camille Troillard, is a particularly effective means of managing Wiimote data (and that of other controllers) by turning it into OSC data. Nintendo stopped production of the Wii (as well as the Wiimote and its successor, the Wii Remote Plus) in 2013.

Gametrak: The Gametrak, released in 2000, was a mechanical controller using two mechanisms beside each other (one for each hand), each consisting of a spool of cable running through a small tube on a ball joint. The small tubes moved like joysticks, sending left and right (x) and up and down (y) data. The amount of cable that is let out determines the depth (z) data to allow positioning of the end of the cable (and whatever is attached). Because of the mechanical nature of the Gametrak and the spooling cable, the Gametrak is particularly well suited to playing music or sound by controlling audio programming environments. As well as the positional data, the interaction with the cable can resemble a musical string, allowing plucking and bowing motions to be mapped onto musical actions.

Leap Motion: While not a specific games controller, the Leap Motion controller was released in 2013 with a focus on virtual reality applications and is a multipurpose input hardware device that uses infrared light-emitting diodes and cameras to identify hand and finger movements.

Other gestural input systems: One of the original electronic instruments is the Theremin, made in the 1920s by Léon Theremin. It operates by sensing the distance between a loop (for volume) and an antenna (for pitch) using capacitance. It remains one of the only instruments played without contact and is a helpful basis for comparison for modern gestural input systems.

The D-beam was a one-dimensional infrared control device incorporated into many Roland keyboards. Its operation was similar to the theremin loop in that the hand's distance above it was measured, but it could be mapped to parameters such as pitch or filter cutoff. The Alexis AirFx, a similar two-dimensional gesture hardware device, is discussed in Section 4.3.4.2.2.

4.3.2.5.3 Wearables
As the name suggests, wearables are items of wearable technology, most commonly in the form of fitness trackers and smartwatches (which can, in some cases, be used as controllers). Several controllers use wearable technology; the most notable are the Mi.Mu gloves, the Genkii wave ring and the XTH Sense.

Mi.Mu gloves: Developed by singer-songwriter Imogen Heap, the Mi.Mu gloves transmit finger flex and wrist orientation data over WiFi. Accompanying the gloves is the software, made using Max, that turns the data from the gloves into either MIDI or OSC, called Glover.

Genki Wave: The Wave, made by Genki Instruments, is an adjustable ring that transmits hand movement as control data over Bluetooth. As well as giving a single button on the side of the ring, the Wave splits movement into five different categories: vertical tilt, horizontal pan, roll (rotation of the hand), side-to-side movement (that they call vibrato), and tap. The software that turns the data from the Wave into MIDI messages is called softwave.

XTH Sense: The XTH Sense, by performance artist and academic Marco Donnarumma, is an open-source wearable that contacts the skin – typically on the arm – and uses two contact microphones to measure the signals produced by muscles. (The origin of the Xth Sense was a crowdfunded project that offered positional and temperature data as well as muscle signals.) This data is then interpreted by software written in PureData (Pd) (see Section 4.4.5). Instead of the Xth Sense being offered as a commercial product, the guide describing how to make the hardware is available online, as is the Pd software.

Smartwatches/smartphones

While smartphones are not typically used as wearables, they can be conveniently strapped to the arm using fitness-based accessories, so they are considered wearables in this context. Many smartwatches and smartphones can sense movement and position through accelerometer and gyroscope data. (Sensor types and descriptions are discussed in Section 4.3.2.5.4.)

Software to manage this data includes MobMuPlat by Daniel Iglesia, made using the programming environment Pd; and the mobile app GyrOSC for iOS.

4.3.2.5.4 Prototyping technologies
All controllers described from Section 4.3.2 to here can be generally summarised as collections of different types of sensors, microprocessors (simplistic

computers that typically run a limited set of repetitive tasks), and software they run to interpret the sensor data. There are several different prototyping options combining sensors, hardware, and software for music technologists with the motivation and time to develop controllers. Two notable prototyping platforms are mentioned here. Sensors and input components are introduced in Section 4.3.2.5.5.

Arduino: One of the most common and flexible prototyping platforms is the range of Arduino microprocessors and the accompanying sensors. Arduino microprocessors are programmed using a variation of the language C++, so they represent a relatively low-level solution, with the corresponding lack of conventions and increased difficulty and flexibility.

iCubeX: Another prototyping solution is the iCubeX system – a range of sensors and interfaces that outputs the sensor data using MIDI messages. Using MIDI and proprietary hardware makes development far quicker but more restrictive and expensive than comparable Arduino sensor-based systems.

4.3.2.5.5 Sensors and input components

The area of sensors and input components is massive, and any thorough discussion is far outside the scope of this book. Accordingly, this section is a brief introduction to the components that form many of the physical control elements of controllers.

Rotary dials are likely to be either a potentiometer (a three-pin variable resistor) for analogue control or an encoder for digital control. Joysticks (including the smaller analogue sticks on joypads) are made up of two potentiometers at 90 degrees giving x and y positions of the joystick.

There is a wide range of buttons or switches that are typically either momentary or latching. 'Momentary' means that the button is only on when held down and will release as soon as the finger is removed. 'Latching' means that the button or switch stays connected after the finger is released and only disconnects when pressed again.

Accelerometers and gyroscopes are sensors that are common in smartphones and smartwatches, as mentioned in Section 4.3.2.5.3.

Accelerometers detect acceleration (e.g., movement from a standstill) in three dimensions. Gyroscopes use gravity to determine orientation, and (unlike accelerometers) can measure rotation. Using accelerometer and gyroscope data together can give a wide range of specific movements that can be used as controller input.

4.3.3 Standalone electronic instruments

Due to the popularity and widespread use of the modern computer, the term 'standalone electronic instruments' is used here to describe hardware devices that do not need a computer to create musical sound. In practice, many of these hardware devices can be used as computer peripherals – typically as MIDI controllers or audio interfaces, but this is only one of many features of these hardware devices.

Despite outputting an electrical signal, electric guitars and bass guitars are not considered electronic instruments because they are not powered electronically.

The manipulation of electric guitar and bass guitar signals is discussed in Section 4.3.4.

4.3.3.1 Keyboards

As with controllers, the most popular type of standalone electronic instrument is the keyboard, traditionally made up of piano keys (typically 25, 49, 61, or 88) alongside a modulation and pitch bend wheel, all of which control an internal sound module. A detailed discussion about the keyboard as a hardware control device is discussed in Section 4.3.2.1.

Different types of keyboards range from stage pianos to workstations and synthesisers.

Stage pianos are designed to replicate a real piano. Consequently, they are wide, have a limited range of sounds and may feature weighted or semi-weighted keys so that pressing the keys feels more like a piano hammer mechanism (rather than synth-action keys, described in Section 4.3.2.1).

Keyboard workstations typically offer an extensive range of sounds and synthesis and sampling sound design methods. Keyboard workstations may also provide several recording studio features, such as multitrack recording and basic editing.

Synthesisers tend to offer a specific range of synthesis types, from specific analogue reissues to modelled synthesisers, and are likely to have synth-action keys.

4.3.3.1.1 Expressive synthesisers

Much like the expressive controllers discussed in Section 4.3.2.4, the following are examples of expressive synthesiser hardware, both of which have an internal sound engine made by Haken Audio called the Eagan Matrix. The Eagan Matrix is a digital modular synthesiser with a comprehensive feature set. Similarly to the Roli synthesisers discussed in Section 4.3.2.4, the Eagan Matrix is designed explicitly for expressive performance.

Haken Audio Continuum Fingerboard: The Haken Audio Continuum Fingerboard was developed by Professor Lippold Haken of the University of Illinois and was commercially released in 1999. It is a long, narrow hardware device with a neoprene surface shaded red and grey in the arrangement of a white and black chromatic keyboard. As well as controlling the Eagan Matrix internal synthesiser, the Continuum Fingerboard can be used as a controller using either MPE or what they refer to as MPE+, a more synchronous and resolute extension of MPE.

Expressive E Osmose: The Osmose, made by Expressive E, is traditional in look and operation to a standard MIDI controller keyboard as it has keys like a piano. The keys have the first three of the control categories discussed in Section 4.3.2.4 and add variations to the continuous downward and upward movement as the keys have a particularly long distance of travel (adding a different area for aftertouch and what they call shake for jerky up and down motion and strum for different sounds triggered over the travel of the key).

The rigid material that makes the played area of the keys means that the Osmose lacks the vertical touch sensitivity of the Seaboard or K-Board Pro expressive MIDI controller keyboards. However, this may mean that transition to the Osmose may be more natural for those unwilling to move away from piano-like keys.

4.3.3.2 Drum machines and samplers

Early drum machines (from the 1980s) used keys or pads arranged in one or more rows to trigger drum sounds from an internal sound module allowing rhythmic sections to be programmed. Notable early drum machines include the Roland TR808, particularly influential in early hip hop, and the LinnDrum, used in numerous popular music hits of the 1980s.

Early samplers, such as the Akai S900, were rack-mounted devices typically triggered by an external MIDI piano keyboard. In the late 1980s, the Akai MPC series of sampling hardware devices was released, using a 4×4 grid of rubber pads to trigger samples. This arrangement was particularly well suited to triggering drum sounds and was highly influential in hip hop and electronic popular music.

Rubber-like pads, often arranged in a grid, have developed into the most common alternative to the piano keyboard for triggering electronic sounds with fingers, especially when their primary use is to trigger percussive sounds.

Modern standalone hardware devices using the 4×4 pad grid approach include the current versions of the Akai MPC and the Pioneer DJS hardware devices. Some, more fully functioned, versions of the Akai MPCs have touchscreens and audio and midi editing functionality like a rhythmically focused DAW. The Pioneer DJS hardware devices are specific DJ samplers designed to work along-side CDJ jog wheels. See the following section for a discussion about turntables and CDJs. It is also worth noting that the 8×8 pad grid Ableton Push 3 can work as a standalone device (versions 2 and 1 are just controllers).

A more current standalone hardware device that uses the drum machine row of keys approach is the Elektron Octatrack, a standalone performance sampler allowing eight audio and midi tracks.

Another standalone hardware device worth mentioning is the Korg Wavedrum – an electronic drum, as the name suggests. It uses sensors under-neath a circular drum head and the rim to trigger sounds from its sound module, which manipulates samples using algorithms specific to the Wavedrum, allowing a distinctive method of performance and a wide range of sounds.

4.3.3.3 Turntables and CDJs

Repurposing the vinyl turntable from a playback hardware device to a highly tactile sound manipulation hardware device and instrument was one of the core developments leading to the creation of early hip hop music. The direct drive design of Technics SL-1200 turntables allowed vinyl disks to be rotated, effect-ively affording manual sound manipulation. The most obvious example of this manipulation is the characteristic rhythmic sound effect of scratching, made by moving the platter (the spinning surface of the turntable) back and forth repeat-edly. Using two or more turntables and a crossfading mixer created two different areas of use: disc jockeys (DJs) and turntablists. DJs mainly playback, loop sections and beat match using the turntables primarily as playback hardware

devices. In contrast, turntablists use techniques like scratching far more extensively, using the turntables and mixer more like an instrument.

Following the vinyl turntable, CDJs were released in the 1990s, using a jog
wheel to replace the platter and physical surface of the vinyl, offering similar
sound manipulation for digital audio. Initially, CDJs were specialised CD players
(as the name suggests); however, in the 2000s, CDJs evolved into controllers for
DJ software, such as the Serato DJ and Traktor by Native Instruments setups.

4.3.3.4 Connection of standalone instruments

Multiple standalone hardware devices are likely to be connected to a mixer. The
most specific example is the DJ mixer, which allows two turntable decks to be
crossfaded and may have other audio inputs that can be mixed in. Depending
on the setup, a mixer is often used as a general balancing hardware device – for
example, in a multiple keyboard setup or as an instrument in its own right, blending
in different sounds, such as in an electroacoustic surround sound installation.

4.3.3.5 Synchronisation of standalone instruments

A synchronisation signal is needed if multiple hardware devices are used
together and are required to stay synchronised. The most common method for
synchronising various hardware devices (including computers) is to use a MIDI
sync clock, connected using either MIDI 5-pin din or USB A-B cables. One piece
of equipment is set up as a host and controls all other connected devices. In this
context, the term 'device' indicates any equipment following the host. It is typically necessary to set the sync setting of any following devices to external sync.
(The MIDI sync clock is not to be confused with MIDI time code, which is a
frame-based series of MIDI messages designed for synchronising with video.)

Link, developed by Ableton, is a synchronisation alternative to the MIDI
sync clock. Link uses the computer networking technology of a local area network to synchronise any computer or standalone equipment connected using
WiFi or Ethernet. Link has no host/device hierarchy, so any hardware device
can control the beat, tempo, and phase, and any connected equipment can leave
or join without causing dropouts when using a MIDI sync clock. While Link is
primarily designed as a synchronisation standard for Ableton Live, several other
hardware and software systems use it.

MIDI sync clock has a long history as a synchronisation standard and is
far more widely implemented than Link. Also, with the development of MIDI
2.0, the new standard does promise more reliable timing than the MIDI 1.0
sync clock.

4.3.4 Effects units

Sound manipulation hardware devices are often called 'outboard' in a recording
studio environment. In the context of a recording studio, these hardware devices
will typically be discussed in terms of 'effects' that are sent (e.g., reverb) and
'processors' that are inserted (e.g., compression). (This is discussed further in

Section 2.) However, in the context of performance and electronic music production, effects and processors are typically referred to using the term 'effects', which is why the broader use of the word 'effects' is used in this book for Chapters 4 and 5.

In a live performance environment, particularly for musicians playing only one instrument, there is a tendency towards a more simplistic insert-only setup. An insert-only setup is likely to be more reliable and allow for quicker problem solving – a consideration that tends to be more critical for live performance events than the recording studio.

A trend explored in Section 5.3 is that the use of effects in live performance events can be a performance element in its own right rather than a sound manipulation technique that takes place in the background.

This section discusses the specific types of effects unit hardware – partly because they are still widely used but also because this sets the contextual framework for software-based effects.

4.3.4.1 Rack mounted

Effects units fall into the categories of hardware devices that are either rack mounted or not rack mounted. Rack-mounted equipment is built to be fitted into a standard 19-inch-wide rack. Each rack-mounted unit has edges that protrude from each side, allowing them to be screwed into the rack, and is manufactured to a height measured in multiples of 'U' so that 1 U is 1.75 inches high. If effects units are rack mounted, this is often an indicator that the equipment is a professional standard (rather than a consumer standard, see Section 1.1.2.2). Rack-mounted equipment is common in the recording studio, for live sound, and to a lesser extent, on stage in live performance events.

The main controls of rack-mounted equipment typically include a set of dials, buttons, and switches and potentially a small liquid crystal display screen – all of which are front facing and designed for manual control. Most inputs and outputs are on the rear, within the rack, and in a recording studio, these inputs and outputs tend to be connected to a patch bay. (A patch bay is a set of rack-mounted panels that allows equipment to be easily connected in a vast range of configurations using patch cables.)

Rack-mounted effects units designed for performance use onstage can often be controlled by foot controllers that are usually made up of several latching switches and/or rocking expression pedals.

4.3.4.2 Not rack mounted

Effects units that are not rack mounted are typically designed to be controlled either by foot or manually. Foot-controlled effect units are commonly called effects pedals, and manually controlled effect units are called tabletop effect units.

4.3.4.2.1 Effects pedals
Effects pedals are designed to be used on the floor and can be split into the general categories of single-switch stomp boxes and larger multi-effect pedals with several switches.

The most common application of effects pedals is the electric guitar (and, to a lesser extent, the bass guitar), where effective use of pedals to alter the tone is often an integral aspect of performance. Guitarists typically create their unique combination of effects pedals mounted on a single pedalboard.

A joint experimental approach is to use guitar effect pedals for other instruments. Electric guitars typically have a mid-heavy frequency range because of how magnetic pickups work, so guitar effects that act like a guitar amplifier will tend to boost the treble and bass components of a signal. While this may produce a pleasing sound in isolation, it is worth being aware that a treble and bass boost (or mid cut and gain increase) is a loudness curve, which simplistically replicates the frequency response of how the typical human ear experiences louder sound. If this curve is overapplied, the music is likely to lack the body and fullness that come from sufficient energy in the mid frequencies.

Another significant application of effects pedals is for live vocals. Vocal effects pedals are designed to be connected to the XLR output of a microphone and are often of a similar form to guitar effects pedals with one or more footswitches.

4.3.4.2.2 Tabletop effect units

Tabletop effect units are designed to be elevated from the floor and mounted on a table or stand. While similar in function to effects pedals because tabletop effect units are intended to be operated by the musician's hands, they typically afford more widely adjustable parameters that can be adjusted live. The most common applications of tabletop effect units are for DJs, vocalists who prefer manual control (especially for beatboxing, a percussive vocal style common in hip hop music), or electronic music performers.

Notable examples of tabletop effect units include the Korg Kaoss pad, which uses a rectangular pad allowing live crossfading of a wide range of effects, and the Boss RC-505 loop station, a tabletop loop station.

4.3.4.3 Loop stations

Chapter 2 discusses the established sound manipulation techniques of the recording studio, covering nearly all of the effects provided by effects units. An effect not discussed in Chapter 2 relates directly to a specific type of performance technique: the loop station.

A loop station is an extension of a delay effect (see Section 2.2.2), which takes a section of audio and repeats it. Delay effects repeat short periods of audio as they happen, with the repetition of the audio reducing (depending on the feedback control).

Loop stations repeat periods of audio – typically from one to several bars long – to create one or several repeating (or looping) and typically musical parts that are often used for accompaniment. A key point about loop stations is that the loops created tend to be fixed once recorded, with no changes in tempo, groove, or dynamics (other than continual layering).

A good early example of fixed live looping is the 2004 KT Tunstall performance of 'Black Horse and the Cherry Tree' on the live music television show

Later with Jools Holland. Tunstall uses an Akai E2 Headrush pedal to layer up a two-bar loop consisting of three rhythmic elements and two backing vocal parts. This loop forms the backing for most of the song until four other vocal loops are added for the final section. Apart from the skill of the performer and the novelty of the approach, this performance works particularly well because it is in a highly repetitive blues style, so it is well suited to fixed live looping. Other notable musicians to use loop stations include Ed Sheeran, Reggie Watts, and Battles.

It is worth noting how using loop stations affects the performance style. When music is performed entirely live (without a loop station or backing track), although the musician may be repeating the same phrase, there will inevitably be variations, however subtle. These variations are likely to be caused by the musician adjusting the expression of their playing in terms of timing or dynamics, or simply making mistakes or chaotic variations (discussed further in Section 5.6). Because the loop being recorded will become all or part of the repeating accompaniment, that loop is likely to be most effective if it is played with as close to a neutral expression and as accurate timing, tuning, and phrasing as possible. Any variation or flair in the loop recording will be repeated and is likely to highlight the potentially mechanical nature of live looping. As well as this, a loop with variation and flair will make the loop less likely to be suitable for layering other parts on top.

Specific live looping techniques are discussed further in Section 5.6.

4.4 SOFTWARE

Several different types of software are of particular interest for musical performance systems. The main software types include operating systems, individual standalone programs, and DAWs with a wide range of music production features, including the running of plugins (e.g., instruments and effects).

4.4.1 Operating systems

The operating system is the general coordinating software that manages the hardware and other software of the computer.

The two most popular operating systems for desktop or laptop computers are Windows, made by Microsoft, and OS X, made by Apple Macintosh. The choice of hardware when selecting a computer is also a choice of operating system.

The OSX operating system is exclusive to Apple Macintosh computers, which are commonly referred to as Macs, so they will be referred to as Macs from this point onwards. In contrast, the Windows operating system can run on computers made by a wide range of companies. These computers are commonly referred to as personal computers (PCs), so they will be referred to as PCs from this point onwards. There are other operating systems, such as Google's Chrome OS for Chromebooks and the open source Linux, but most audio programs, such as DAWs, are only made for Macs or PCs.

PCs have a far larger market share than Macs because they typically have better raw computing specifications for the price, making them more commonly associated with general office work such as word processing and web browsing.

In contrast, Macs have long been associated with high-end creative situations such as graphic design and audio. In broad and very general terms – partly because Apple has more control over the hardware and the operating system – Macs have developed a reputation for being more reliable than PCs. However, Macs have also developed a reputation for being less flexible, more expensive, and having an even higher degree of designed obsolescence than PCs. Commercial audio software made to work on Macs and PCs is called 'cross-platform'.

4.4.2 DAWs

DAWs are the culmination of over a century of audio and music technology – in particular, the evolution of music production techniques associated with the recording studio. The most fundamental examples of these techniques include recording and editing multitrack audio and signal processing (i.e., the processors and effects discussed in Chapter 2). DAWs also make use of the sequencing capabilities of MIDI, as well as the sound creation and manipulation techniques of samplers and synthesisers. DAWs are the most widely used technology for modern music production and a powerful composition, collaboration, and performance tool.

Important aspects of any system are its affordances and – of particular interest to an overview of DAWs – the degree of procedurality.

Affordances: To briefly summarise, 'affordances' are what something indicates to the user that it can do. To quote the design researcher Don Norman[3]: "affordance refers to the perceived and actual properties of the thing, primarily those fundamental properties that determine just how the thing could be used." He continues to give the examples: "Knobs are for turning. Slots are for inserting things into. Balls are for throwing or bouncing" (p.9).

Procedurality: 'Procedurality' is a term that I use to describe how limited a system is in its methods of operation. A system that only allows one way to record and edit audio is highly procedural. Another system that allows a wide range of ways to record and edit audio is far less procedural. Although both offer a similar approach to music production, the DAW Pro Tools, for example, is generally more procedural than the DAW Reaper. The more procedural design of Pro Tools gives it a more standardised workflow, so it is easier for sound engineers to move from one studio to the next and work to a reliable and commercially honed music production workflow. The less procedural design of Reaper makes the workflow far more customisable and open to a more personalised and experimental workflow. Affordances and procedurality are explored and unpacked further with a more theoretical framework in my forthcoming academic book on liveness in modern music performance.

For an approach that focuses on the use of technology for live musical performance, it can be helpful to separate DAWs into DAWs that have the facility for non-linear clip launching and those that do not. For the sake of brevity, DAWs that do not have the facility for non-linear clip launching are referred to here as 'linear timeline DAWs', although non-linear clip launching DAWs are also likely to have a linear timeline mode.

4.4.2.1 Linear timeline DAWs

The traditional and typical mode of operation for DAWs is to follow a linear timeline with a main window (called the edit window in Pro Tools and arrangement view in Ableton Live). This main timeline window serves as a graphical representation of a multi-track recorder that shows tracks stacked vertically and the audio or MIDI arranged in a timeline from left to right, with options to show additional information such as automation, fade-ins, and fade-outs. When the transport for a linear timeline DAW is started, the clips play linearly, deviating from that mode only if a loop or repeat section is active (causing a section to repeat until transport is stopped). Figure 4.9 shows a small section of the edit window of a standard multi-track session in Pro Tools.

Linear timeline DAWs are so well suited to traditional offline production processes that they have become the almost exclusive tool for the task. This popularity has meant that linear timeline DAWs greatly influence other processes, such as music composition, outside of what would be expected to be electronic genres.

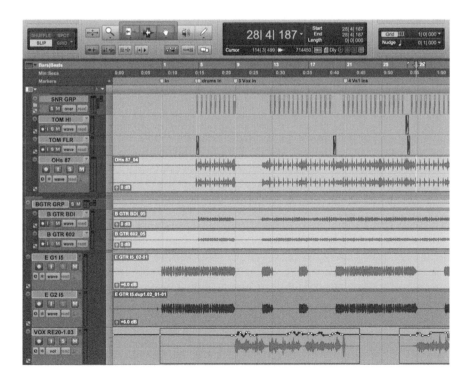

FIGURE 4.9
Edit window in Pro Tools

4.4.2.1.1 Eager linearisation and visualisation

Creating media on a linear timeline is likely to be more procedural than creating media using non-linear clip launching. The linear timeline is a system that encourages what Duignan, Noble, and Biddle[4] refer to as "eager linearisation" (p.726).

Music composition can be considered the linear organisation of musical events, or linearisation. Therefore, systems that employ eager linearisation encourage committing this organisation earlier in the creative process. Linear timeline DAWs demonstrate particularly eager linearisation as the ordering of events is central to the early stages of their typical workflow. A linear timeline also affords a more procedural and, therefore, streamlined and reliable approach, which is why this is also the general method of mainstream media post-production. This eager linearisation does, however, limit the opportunities for spontaneity, both to produce recorded music for release and, more obviously, to produce music to be used in live performance events (e.g., accompaniment, including backing tracks). In contrast, the most delayed form of linearisation would be a wholly improvised performance. However, the conventions and form of the musical genre would provide a framework for linearisation at the time of performance.

Another factor that significantly influences modern music production is the amount of visual editing that DAWs afford. In the analogue recording studio, music visualisations are typically limited to LED and VU needle type level metering, giving little choice other than to listen intently. In the modern DAW environment, nearly everything is visualised, including audio waveforms, MIDI notes, the associated clips, automation, and spectrum analysers. Accordingly, electronic music tends to be produced in a more visual rather than aural fashion. For example, in an article in the *Guardian*, music producer and singer-songwriter James Blake[5] expressed the appeal of producing songs visually using the then solely linear timeline DAW Logic:

> I could record them and look at them, almost physically – graphically – and just chop up what I did like and I didn't like. … It didn't have to be all in one take, it could be something I designed from the ground up, visually.

Visual editing is a logical workflow for music that is built more than recorded (see Section 2 for a discussion of these processes). However, it is likely that as performances are more heavily edited and 'fixed,' there is potentially less value in retaining an expressive performance if it can be transformed into a perfectly consistent part. Fortunately, there is a choice in workflow and the degree to which music is recorded, built, enhanced, and edited. It is likely that more expressive music resulting in a combination of highly skilled performance and production processes will have a uniqueness that would otherwise be impossible to produce.

4.4.2.1.2 Pro Tools and Cubase

There are several different linear timeline DAWs, each with their own strengths and weaknesses, and it is beyond the scope of this book to discuss them all. The linear timeline DAWs discussed here are Pro Tools and Cubase, chosen for their popularity.

Pro Tools, made by US company Avid, is a good example of a DAW designed to resemble the workflow of a physical recording studio. Pro Tools has its roots in digital audio recording and is often integrated with accompanying Avid recording studio hardware (i.e., microphone preamps and converters). Accordingly, Pro Tools has gained a reputation for being the industry standard for rock and pop music production.

Cubase, made by German company Steinberg, has its roots in MIDI sequencing, and Steinberg was the company to develop virtual studio effects and instruments. Accordingly, Cubase is an example of a DAW with a more flexible and electronic music production approach, evidenced by its MIDI drum editor, a specific editor optimised for more convenient programming and editing beats than the piano roll editor.

4.4.2.2 Non-linear clip launching DAWs

Non-linear clip launching DAWs tend to have a linear timeline, but they also have another mode of operation affording the non-linear launching of clips. The first major DAW to afford non-linear clip launching was Ableton Live.

This mode of operation (called session view in Ableton Live) allows clips to be launched (and therefore ordered) at the time of performance either by manual launching (typically using pads or keys) or by a set of behaviours. While the clips themselves are created before the performance, in comparison to a linear timeline DAW, non-linear clip launching demonstrates a lower level of procedurality and more delayed linearisation.

Non-linear clip launching DAWs are the significant mainstream alternative to a linear timeline approach. As mentioned previously, the first DAW that offered non-linear clip launching was Ableton Live; so, much like Pro Tools for traditional music production, Ableton Live has become the standard for non-linear clip launching. While initially this approach was restricted to a DJ-style setup, the popularity of a non-linear approach has caused newer DAWs to be created with non-linear clip launching as a central feature (e.g., Bitwig). More traditional DAWs have also been updated to include non-linear clip launching functions (e.g., Digital Performer and Logic).

Non-linear clip launchers typically use a grid of clips (rather than the linear timeline representation of multi-track recording) that are launched either individually or in groups. If the tracks are arranged in vertical columns, the rows of adjacent clips can be launched together and are typically called scenes. Figure 4.10 shows ten tracks of the non-linear clip launching mode of Live – the session view. Ableton Live has two main modes of operation: the clip launcher (session view) and the linear timeline (arrangement view). These two modes can be used together, with some tracks running an arrangement and others launching clips from the session view. The incorporation of the programming environment Max for Live in Ableton Live is notable, particularly as it has access to the Ableton Live application program interface (API). Max for Live is discussed further in Section 4.4.5 and is the environment I used to create the reactive backing plugins discussed in Chapter 5.

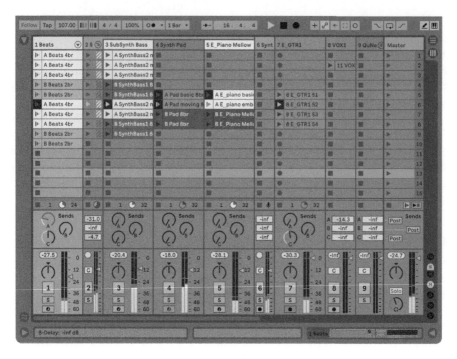

FIGURE 4.10

Session view in Ableton Live

Image Line's FL Studio (previously called Fruity Loops) is another DAW with its roots in DJing. FL Studio was initially focused on loop-based music and has grown to encompass a wide range of electronic music production techniques. The clip launcher mode of FL Studio is called performer mode. This approach clearly divides the use of the software between the offline linear timeline operation and a non-linear clip launching mode optimised for DJ-style performance.

4.4.3 Plugin hosts

While DAWs can be used for live performance (e.g., Ableton Live), they include a wide range of feature sets that take up valuable processing capacity and add system latency. Using a DAW also encourages an extended form of performance – for example, the augmented performance techniques discussed in Chapter 5.

If these features are unnecessary, and all that a DAW is used for is to manage plugin instruments, then a plugin host may be a more practical choice. Plugin hosts are software programs that focus on the smooth management of plugin instruments and effects and the playing of backing tracks and/or click tracks without the deep MIDI and audio clip functionality of a DAW.

Plugin hosts run a setup of electronic instruments, playable in a more traditional fashion, where the computer is used as a highly flexible sound module. In practice, plugin hosts switch between different configurations of instrument patches and effect presets as smoothly as possible, with as little risk of crashing as possible.

Examples of plugin hosts include Mainstage, made by Apple (Mac only), VST Live, by Steinberg (Mac and PC), Gig Performer, by Deskew Technology (Mac and PC), and Cantible, by Topten Software (PC only).

4.4.4 Software instruments

Software instruments are computer programs that create musical sounds, typically using synthesis and sampling techniques. While many software instruments can run as standalone programs, it is more common to run instruments as plugins within either a DAW or plugin host.

A standalone program will allow the instrument to be played. It will manage incoming and outgoing audio and MIDI, but switching between different standalone software instruments is likely to be highly inconvenient.

In a DAW, software instrument plugins are typically inserted onto a track and then managed using the functions of the DAW, either playing music by triggering MIDI clips or playing the plugin live.

In a plugin host, software instrument plugins are typically managed as different patches configured for a live performance scenario.

There are various plugin formats, the first and most widely compatible being .vst, created by Steinberg. Other notable plugin instrument formats include the Apple-only .au and the Pro Tools-only .aax.

All these plugins are triggered by MIDI 1.0 messages, some are compatible with MPE, and soon plugins are likely to offer the expanded capabilities offered by MIDI 2.0 (MPE and MIDI 2.0 are explained in Sections 4.2.4 and 4.2.5 respectively). A notable addition is the Steinberg VST3 format, which also expands on the MIDI 1.0 messaging conventions discussed in Section 4.2.

4.4.5 Audio programming environments

As the name suggests, audio programming environments allow the creation of audio creation and manipulation software devices – most notably Max by Cycling 74 (at the time of publishing, Cycling 74 was owned by Ableton), and Reaktor by Native Instruments.

Some audio programming environments are text-based, such as Supercollider and ChucK. Max and Reaktor, however, are high-level programming environments, using visual chunks of code that are patched together with lines representing the flow of information. These lines and chunks of code can be seen in Figure 4.11, which shows an example of a Max patch – specifically, a subpatch that manages clip slot recording and playback in the RB.sdlooper2×3 device running in Max for Live.

The information flow in Max is top-down (the large boxes with 1 and 2 at the top of the patcher window are the patched inputs). In Reaktor, however, the information flow is left to right.

A free to download variation of Max is the open source software Pd.

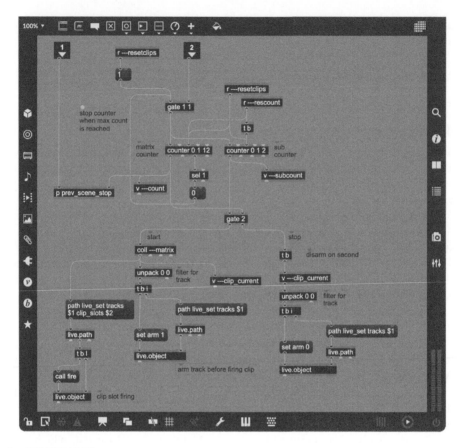

FIGURE 4.11
Max patcher window

A variation of Max that works as a part of the DAW Ableton Live is called Max for Live, shown in Figure 4.11, and as mentioned in Section 4.4.2.2, this integrates with Ableton Live and its API. Max for Live is the environment I used to create the devices discussed in Chapter 5. Figure 4.11 is also shown in Section 5.4.4.2, where it is explained. A specific set of thorough tutorials for reactive API use of Max for Live is outside the scope of this book, but please feel free to contact me if this would be useful. If there is significant demand, I will make these available online.

4.5 AUDIO FOR IMPORT

One of the primary uses of audio software such as DAWs is to record (and then manipulate) sound either in the recording studio or outside (referred to as field recording or location recording). The alternative to recording sound is to import pre-produced audio. This audio is typically referred to as samples. (The use of samplers and samples is discussed in Section 2.3.6b.)

4.5.1 Commercial releases

Creative works are not made in a temporal vacuum; they are informed and inspired by previous work, regularly borrowing and repurposing elements from previous work. The electronic technology of the mid-twentieth century led to the creation of a wealth of devices affording direct insertion and manipulation of commercially released musical recordings. In turn, these devices pushed the creation of many new types of music.

In the art music world, this initially resulted in forms of tape music, where the composers were likely to have created the recordings themselves.

However, in popular music, it became popular to manually loop sections of previously recorded music on vinyl disks by DJs, forming the musical basis of hip hop and rap. At first, this was a process of using two of the same disks containing a musical break (i.e., an interesting rhythmic section with little lead parts such as singing) that could be played on one disk while the other was spun back to start over.

As digital technology progressed, this led to the creation of samplers, which made looping much easier to duplicate and encouraged further audio manipulation. As well as these looping techniques being fundamentally responsible for hip hop and rap, digital sampling (as well as developments in synthesis) led to an extensive range of popular electronic music genres and influenced nearly all other popular music genres.

Several subgenres of hip hop and rap were driven by producers searching for sample-suitable commercially released vinyl disks or 'crate digging'. However, this process was severely limited by several high-profile legal cases – most notably when singer songwriter Gilbert O'Sullivan successfully sued rapper Biz Markie in 1991 for sampling the song 'Alone Again (Naturally)'. This case set a significant precedent that made the process of sample clearance (i.e., acquiring permission to use the sample from both the songwriter and the mechanical copyright holder) a vital part of the commercial release process.

Using sections of commercially released popular music as samples to create more music results in audio that is, in part, a result of the production processes of that time. The rudiments of the techniques used for mixing and mastering are discussed in Chapter 2. These techniques can be summarised as a combination of editing, equalisation, compression, and reverb, at the very least. Accordingly, a significant difference between produced and raw recorded audio can be summarised as follows. Produced audio is likely to have a far lower dynamic range and far more sound energy across frequency bands (caused by compression, limiting, and maximising) than raw recorded audio.

Often the point of sampling commercially released music is that the character of the original recording is inserted into a new piece of music so that it will intentionally stick out as a reference or feature. If this is not the case (or sometimes even if it is), making that mixed and mastered audio sit effectively into the new mix may be challenging. Further manipulation of the sample may be necessary – for example, expansion, equalisation, or dialogue de-reverb on a vocal sample so that a reverb suitable for the new production aesthetic can be applied.

4.5.2 Sample packs

The other type of audio that is likely to be imported for music production is specifically produced sample packs (or loop packs). Sample packs are collections of audio files to be imported and manipulated to produce musical tracks. The use of sample packs is common in electronic dance music and pop music. For example, the rhythmic elements of the 2007 pop hit 'Umbrella' by Rihanna (feat. Jay-Z) are almost exclusively the loop 'Vintage Funk Kit 03' from the free Apple DAW Garage band.

Many DAWs will include libraries of one-shot samples and phrase loops; and there are many producers of sample packs that can be purchased, allowing a quick and convenient Lego-like assembly of music. However, music produced predominantly using unaltered audio from a sample pack is likely to sound quite derivative and lack originality. As with most audio, sample packs can be played as audio or triggered by a sampler, and many will come pre-configured for well-known samplers and sample players.

Similarly to samples taken from commercial popular music releases, samples from sample packs are likely to have been manipulated to sound bigger and brighter – that is, to have a lower dynamic range and have more sound energy across frequency bands. However, as they are not a part of a piece of music, samples are also likely to have been normalised to something close to 0 decibels relative to full scale (dBFS) (unless the samples are lower-velocity layers), as audio with higher gain tends to sound more immediately impressive. Normalisation is a two-stage process of analysing audio to find the peak level and then increasing (or decreasing) the gain of the entire audio so that the peak level reaches a specific level, such as -0.3 dBFS. It is good practice to reserve normalising audio until the final mastering stage, as this is likely to help maintain good gain staging.

Accordingly, the track level that the samples are on is likely to be higher than the tracks with less processed audio. Often, sample tracks will need to have their gain lowered to maintain headroom and avoid unwanted overdriving and clipping in bus devices. Specifically, this means that good practice will be to keep maximum track levels around -15 dBFS on each track depending on track count, as more tracks in use will sum to give a higher signal level. A common mistake many make when starting to mix is to keep moving faders up to balance, which quickly takes out the headroom and creates unwanted overdrive and clipping in bus devices. It is often better to get into the practice of turning other tracks down (either individually or in groups) to achieve a better balance.

ENDNOTES

1.   The MIDI Manufacturers Association. (2019). *The MIDI Manufacturers Association (MMA) and the Association of Music Electronics Industry (AMEI) announce MIDI 2.0™ Prototyping*. [Blog]. MIDI Association. https://www.midi.org/articles-old/the-midi-manufacturers-association-mma-and-the-association-of-music-electronics-industry-amei-announce-midi-2-0tm-prototyping

2.   Sonicstate. (2017). *Tim Exile and his Flow Machine* [Interview]. https://youtu.be/4aNyhKcHdts

3.   Norman, D. (1988). *Psychology Of Everyday Things*. Basic Books.
4.   Duignan, M., Noble, J., & Biddle, R. (2005). A Taxonomy of Sequencer User-Interfaces. *Proceedings of the International Computer Music Conference*, 725–728.
5.   Needham A. (2011). 'James Blake: "I didn't make this record for Chris Moyles, I'm in the dubstep scene."' *The Guardian*, January 22. https://www.theguardian.com/music/2011/jan/22/james-blake-dubstep-scene

Music technology in live performance

Τhis chapter focuses on innovative and experimental systems and techniques to augment live musical performance events. Many of these systems and techniques attempt to move away from fixed and independent methods of accompaniment that limit the effective exchange of energy, such as backing tracks, and towards systems that use reactive, interactive, and chance elements.

There are many established systems for augmenting live music performance events (the main system types are discussed in the previous chapters). This chapter deals with system elements relatively new to a performance environment. New and experimental approaches are mainly untested and impossible to examine within established conventions because they are no longer experimental if they become conventional. This chapter discusses these systems and approaches with the understanding that they are under development, which is part of what makes them so exciting but also so difficult to predict.

Before exploring these technologies and techniques in Section 5.2 onwards, underpinning concepts are introduced in Section 5.1. For a more extensive theoretical exploration of these concepts, see my upcoming book on liveness in modern music performance. To skip to the specific technologies and techniques, go to Section 5.2 and refer back to the particular concepts as needed. Otherwise, read on…

5.1 UNDERPINNING CONCEPTS

'Creative work' is defined here as any activity working towards an innovative output. This definition is intentionally broad to avoid the cultural boundaries of work considered 'art', 'craft,' or 'product'.

For creative work and the creation of devices for creative work (in itself creative work), an understanding of the context in which the work is taking place is needed – for example, the relevant genre conventions and inspirational precedents that convey meaning. Without this, successful and effective work is just a fluke, and any meaningful creative development will be limited. To put it another way: no work is created in a cultural vacuum, and accordingly, greater awareness of the cultural context gives greater control over creative resources.

 DOI: 10.4324/9781003370406-6

5.1.1 Boundaries of originality and effectiveness

The challenge of objectively measuring the effectiveness of any creative work is substantial and changes as culture does. It is taken here that the desired effect of creative work is likely to be a meaningful exchange between the work and people. Effectiveness can be determined by how meaningful that exchange is for how many people and for how long. While this definition may resemble a long-term popularity contest, this measure is also guided by places that offer a degree of cultural quality control, such as art galleries, record labels, and universities. However, these institutions are by no means immune to their own or more systemic forms of bias.

Marking out the boundaries of effective work is possible when we identify the characteristics of work broadly accepted as *not* being effective. Identifying these characteristics can help us avoid these pitfalls in our work, recognise them in other work, and build genuinely innovative tools.

There are broad cultural boundaries of what counts as creative work that is not effective. Work that is not effective can be summarised as work that represents two extremes of a uniqueness scale – that is, which is either:

• not unique at all – overly derivative of other work in too many elements, or
• far too unique – overly obscure and unrelated to any recognisable work in too many elements.

The first of these types – the not unique at all, overly derivative type – is work that is purely a copy of elements from other works, adding nothing original. The overly derivative type may be fine for beginning exercises, where the creator is learning techniques and the craft. However, for work to be successful when released, performed, or exhibited, it needs a degree of originality: its own creative 'voice'.

The second of these types – the far too unique, overly obscure type – is work that demonstrates no discernible control over creative resources – musical, production, technical, or otherwise. Because too many elements do not fit any particular form, meaning or expression is not communicated.

These two types are shown in Figure 5.1 as the two extreme points on a scale of uniqueness that move away from a zone in the middle. This 'zone of effective work' is effective because it has a combination of elements that are both unique and make recognisable and flexible use of form. The figure uses images for facial recognition as a metaphor for this point. On the left-hand side of Figure 5.1, all four of the faces are identical, so there are no differences between them, and no faces are individually recognisable (the faces are also intentionally formulaic as they are derived from emojis.) On the right-hand side of Figure 5.1 is a jumble of lines, all different but with no recognisable form of faces or anything else.

It is worth noting that work that is more unique and outside of conventional form may be simply ahead of its time, but we can only judge based on what we know at the time.

For technology development, it is worth a regular check-in to ensure that the technology does not encourage work that is either overly derivative (i.e., no uniqueness) or too obscure (i.e., no apparent use of form). For example, a

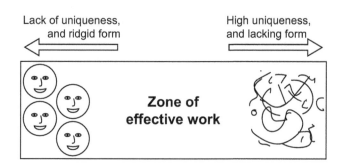

Lack of uniqueness, and ridgid form

High uniqueness, and lacking form

Zone of effective work

FIGURE 5.1

Model of effective work based on uniqueness and form

common trope of digital audio technology is an over-reliance on skeuomorphism in the graphical user interface (GUI). Skeuomorphism is the designing of GUI elements that mimic real-world counterparts. Skeuomorphism can make controls a familiar experience – for example, a filled circle with a number underneath and a dot indicating its value signifies a dial. It is common practice that a mouse clicked on that circle can raise or lower that value in the same way a physical dial would. While a simple number box would be a more efficient use of the GUI, it is generally much easier to gain a quick overview of a set of controls by glancing at a familiar set of dials and sliders.

However, skeuomorphism can result in an overly derivative user interface of a standard set of devices. The reliance on adhering to physical device standards (as well as any general well-trodden standards) may be a distracting and limiting factor. For example, the controls to an 1176 compressor (see Section 3.1.2.2) are a result of the design limitations of the time and work particularly well for that specific example. However, a surprising number of compressor plugins unnecessarily model these limitations, unnecessarily restricting device flexibility. The number of these backwards-looking effect recreations demonstrates the appeal of nostalgia and the path dependency mentioned in Section 4.3.2.4.

On top of this, lots of computer processing may be taken up, generating depth or physical material effects which have no bearing on the audio processing. An example of this is the first version of the Native Instruments synthesiser plugin Massive X, which had a user interface that was particularly processor intensive. While an excellent synthesiser plugin, this meant that the synthesiser would not be likely to be able to be used in a live set where processor headroom was limited. Fortunately, later versions introduced an option to flatten the user interface, making the synthesiser work far more efficiently.

5.1.2 'Should' factors

While in any area of creative work there is likely to be an ebb and flow of trends, technological influences bring the allure of shiny and new technologies and techniques. It may be human nature to desire new things and ways to use them, but this desire has been weaponised by centuries of dogmatic teaching and decades of highly effective marketing. This undue pressure can obscure rather than augment creative practice and often takes the form of what is refered to here as 'should' factors.

It may be tempting to use technologies and techniques *solely* because they are available, new, or novel. However, without input that has a meaningful aesthetic purpose, the work is purely a gimmick – a vehicle to show off the technology or technique – and is very unlikely to be effective.

An in-depth discussion of the subjective and objective aspects of aesthetics is outside the scope of this book. However, the measure of 'effectiveness' is chosen (rather than a more arbitrary and fixed concept of 'quality') as the 'effect' of creative work depends on the intentions of the creators and the receptions of society, which offers a more flexible and inclusive measure.

Creating work with a high chance of being effective is driven by a skilled and confident process with an aesthetic purpose underpinned by critical self-awareness. Common elements that can detract from this process include ones chosen not to serve the aesthetic but because the creator is unduly influenced by factors that are arbitrary to the aesthetic purpose, rigid, and chosen mainly because the creator feels they should. These are referred to as 'should' factors from this point onwards. The word 'should' is placed in inverted commas to indicate the restricted and confidence-sapping interpretation of the word that can obscure creative flow and a sense of play. 'Should' factors are the opposite of critically applied best practices. They tend to be dogmatic and are caused by a more insecure interpretation of the technologies and techniques that 'should' be used in creative work.

There are two types of typical 'should' factors:

- sales hype, which manages to convince that work can only be good if a particular device or proprietary technique is used
- a learned narrative which asserts that the creator *must* use a particular technique or device for the work to be good, whatever the situation.

Any inclusion of these 'should' factors is likely to inhibit originality and effectiveness, but this is not a binary division – it is a process of critical awareness. (If any 'should' factors are apparent in this book, please let me know!)

The lines between useful new product features and pointless gimmicks or finely honed craft and dogmatic application of technique can be blurred. However, a quick test can be made by identifying clear and critical aesthetic purposes or goals of the work and asking the question: "Does this technique serve the creative/aesthetic goals of the work?"

This question is suggested as a core part of an effective and reflective creative process and is integral to the techniques discussed here.

5.1.3 Concepts underpinning liveness

The liveness of a musical performance event describes, as the name suggests, how live that event is. The concept of liveness exists in contrast to the fixed nature of recorded audio. Any recorded audio or fixed music (e.g., a fixed Musical Instrument Digital Interface (MIDI) clip) is referred to here in the context of live performance as 'non-live' musical elements.

Liveness is particularly relevant when we consider how culturally dominant recorded audio is, from the old music consumption formats of CDs to streaming platforms and broadcast media such as television and social media. Recorded music is listened to far more than live music because of the relative convenience, cost-effectiveness, and ubiquity of recorded media.

One crucial point is that using non-live elements in live performance events tends to reduce its liveness. For example, a group playing along to a fixed backing track has significantly less liveness than another playing exclusively acoustic instruments. The group playing with the fixed backing track has to follow the backing track's tempo and structure, leaving little room for spontaneous tempo, groove, and arrangement variation. Along with this reduced liveness comes an interference with musical cause and effect and exchange of energy because not all music heard is being played live.

5.1.3.1 Practical conditions of performance liveness

Oxford Dictionaries defines 'liveness' as 'The quality or condition (of an event, performance, etc.) of being heard, watched, or broadcast at the time of occurrence'.

This definition sets some rudimentary boundaries – most notably that the performance occurs *at that time*, so the performance is unique to that moment. However, in successful live musical performance events, the music does not just occur; it is designed, rehearsed, and skilfully performed. Using this expanded definition gives two practical conditions for musical liveness in the context of the performer. A musical performance event has an amount of liveness that depends on two things:

1. how **unique** that performance is to the moment, and
2. how musically **expressive** that performance is.

5.1.3.2 Causation

A vital element of a live musical performance event experience is the belief that what is being watched is the cause of the music being made, and that the movements made by the musicians are an integral part of the act of creating this music. For example, a singer may move their hands outwards when singing a phrase. Although technically, this has little practical effect on the sound of the voice, the expression associated with reaching out is clear. The movement combines with the singing to heighten the expressiveness of the performance. Also, when a singer is singing live, the singer and their mouth, throat and torso movements are the direct cause of the vocal sounds, and the rest of the body movements are likely to be indirectly causal to the expression of the singing.

When non-live elements are used, the cause and effect of music production are often less clear. A specific and extreme example of this would be in 1992 when popular electronic musicians The Orb played chess on the UK TV show *Top of the Pops* while their single played. This overt example of pressing play of pre-recorded material makes an interesting statement about the development of electronic dance music, but it is not a musical performance.

However, it is not simply the case that musical performance events are either causal or not causal. Electronic technology has led to three categories of causation: direct, indirect, and virtual causation.

Direct causation: Direct causation is when the musician is the primary cause of any ongoing variation of each note's musical pitch, duration, dynamics, and timbre in real-time. For example, a musician playing the saxophone is directly causing note characteristics by their ongoing mouth shape, breathing, and finger movements. Direct causation is the ingrained musical situation in place for centuries because of how acoustic musical instruments are played.

Indirect causation: Indirect causation is when the musician is causing the music to be created, but a significant proportion of the ongoing pitch, dynamics, and timbre variation of each note is caused by a system – for example, when launching a phrase sample.

The critical factor here is that one or more systems (typically digital ones, such as modulation sources) have significant control of the note's ongoing pitch, dynamics, and timbre. The playing of electronic instruments is likely to include some degree of indirect causation.

Virtual causation: Virtual causation is when an alternate and fictional narrative is offered for the cause of the music – for example, the cartoon characters of art-pop band Gorillaz. While the cartoon characters are not actually playing the instruments (sorry if this is a spoiler), the fiction of this virtual band is clearly appealing, as seen in the band's enduring popularity. While virtual causation cannot be live in any way that authentically communicates human agency (different kinds of agency are discussed in Section 5.1.3.5), it offers a range of possibilities to communicate musical meaning in innovative ways.

5.1.3.3 Musical-visual agency

The term 'agency' is the capacity to act in a way that produces a particular result. In video game design, for example, a character's agency refers to the choices available to that character.

'Musical-visual agency' is the visual demonstration of the specific action(s) that produce the particular musical result(s). In this context, a human agent is a musician, and a machine agent can be anything from a piano keyboard to a reactive and chance-based digital audio workstation (DAW) (techniques discussed in Sections 5.5 and 5.6, respectively).

A helpful parallel here is the use of machine agents (i.e., computers) in chess. The victory of the computer system Deep Blue against Garry Kasparov in 1997 is seen as a milestone in machine agency, demonstrating computer dominance over humans in chess. It is now widely accepted that a 'centaur' – the combination of a skilled human and suitable software – is undoubtedly the best at chess,[1] giving an unbeatable combination of brute force tactics and meaningful strategy.

In musical performance events, it is the 'musical centaurs' – the combination and interaction between humans and machine agents – that offer such novel possibilities for the future. However, there are many potential opportunities for faking human musical-visual agency – whether the 'press play' sets of the DJ world or the backing tracks from other forms of popular music. Making sure that the visible actions (or performance signifiers, see Section 5.1.3.5) are perceived as authentic (as in the case of musicians playing known acoustic instruments) is vital to human musical-visual agency.

Human musical-visual agency is demonstrated in playing most acoustic instruments, where the human agency creates the direct causation discussed in Section 5.1.3.2. This human musical-visual agency becomes increasingly more nuanced with culturally well-known instruments, such as the electric guitar in rock music.

Machine agency is somewhat more controversial. Agency is often reserved only for humans, as machines (currently) cannot show consciousness or intent. However, a broader interpretation of agency gives us the opportunity to bring machine processes into a more complete analysis of modern music performance events.

From the perspective of a traditional guild approach, machine agency is a clear threat to the old order. Godlovitch[2] sums up resistance to computer interference in music with a painting analogy that illustrates both the reductive fear and misunderstanding of digital tools: "Computers produce code, not colour" (p.100).

Of course, computer code *does* produce colour, only with a different kind of palette. That the computer uses a numerical representation for colour seems to be at the heart of the fear and distrust of computer processes. This fear arises from the assumption that our human understanding of colour, sound, or anything is completely accurate and 'real' when what we experience – that is, our reality – is itself a representation.

While resistance to machine agency is entirely understandable from those arguing against faking musical performance at live events, this understandably controversial topic represents only a tiny part of technological influence.

5.1.3.4 Live, non-live, and pseudo-live musical elements

When considering the specific elements of live performance events, these elements can be split into three different areas: the live, the non-live, and the pseudo-live. These are shown in the columns of Figure 5.2.

Live musical elements: Live musical elements have a significant amount of direct causation in their performance. As described in Section 5.1.3.2, direct causation is when the musician is the primary cause of any ongoing variation in musical pitch, duration, dynamics, and timbre of each note, in real time. Most acoustic instruments demonstrate direct causation, so images of a violin and piano are typical examples of instruments that count as live musical elements.

Non-live musical elements: Non-live musical elements lack any significant direct causation after their launching – that is, the performer has no influence, direct or otherwise, over each note's ongoing musical pitch, duration, dynamics, and timbre. Common examples of non-live elements include fixed backing tracks, phrase samples or MIDI clips that play independently of the music.

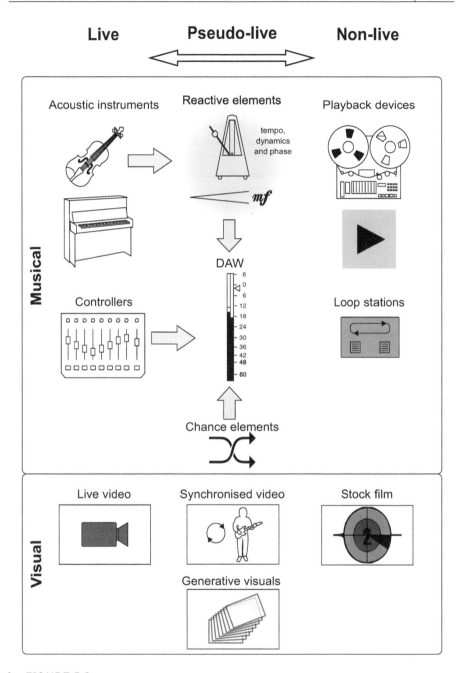

FIGURE 5.2

A model of live, pseudo-live and non-live musical elements

Pseudo-live musical elements: Pseudo-live musical elements are the specific musical elements that sit in the middle ground between the live elements (with significant amounts of direct causation) and the non-live musical elements (which lack any substantial direct causation). These elements have indirect causation – that is, they are systems (typically digital) that control a significant proportion of the ongoing pitch, dynamics, and timbre variation of each note independent of the musician. This indirect causation is likely to come from either reactive elements (explored in Section 5.5) or chance elements (explored in Section 5.6).

There may be a certain amount of direct causation – that is, a musician launches a phrase sample and adjusts one macro parameter of the modulation, but not necessarily. The essential factor for pseudo-live musical elements is that they partially satisfy at least the first of the practical conditions for liveness – that it is unique to that moment; and, if possible, also the second – that it is musically expressive.

5.1.3.5 Performance signifiers and musical-cultural signifieds

There are many musical events, from club nights and gigs to art installations and orchestral concerts; however, the use of non-live musical elements raises the question: **when does a musical event stop being a performance?**

In music performance, skill and control need to be communicated to allow musical expression and exchange of energy. The communication of this skill and control is referred to here as 'human musical-visual agency'. **A musical event stops being a live performance when it has no significant human musical-visual agency.** For example, as described in Section 5.1.3.2, The Orb playing chess on the UK TV show *Top of the Pops* while their single played is clearly not a musical performance.

We can dig deeper into the component parts of human musical-visual agency to better understand musical performance and to aid in creating tools for performance.

In the model in Figure 5.3, human musical-visual agency is shown as a combination of causation, performance signifiers, and musical-cultural signifieds.

Causation, and the different types of causation relevant to musical performance events, are introduced in Section 5.1.3.2. Along with causation, performance signifiers are the components caused by the performers, and in contrast, the musical-cultural signifieds are experienced by the audience. Causation in live musical performance events is explored in Section 1.1.4. Direct causation is likely to be more causal than indirect causation; however, either type of causation counts.

The terms 'performance signifiers' and 'musical-cultural signifieds' are variations of concepts taken from the field of semiotics. Semiotics is the study of signs and symbols, their meanings, and what and how they communicate. A sign is a specific and descriptive form of language – for example, a give way road sign. A symbol is a conventional representation, likely to have a less specific meaning – for example, a Christian cross. A musical performance event has many visual features that are highly symbolic.

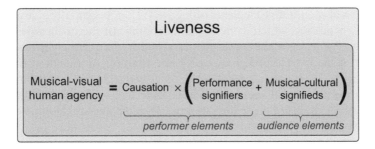

FIGURE 5.3

Components and formula of human musical-visual agency as an element of liveness

In semiotics, a sign or symbol comprises two elements: a signifier and a signified. The signifier is the part that can be perceived, which indicates what it is – for example, the words 'GIVE WAY' within a red triangle pointing down. The signified is the concept – for example, the rule that cars from the other direction have priority and the driver must stop to let them pass.

In musical performance events, the performance signifiers are the combination of the music and any visual indicator of musical performance actions – for example, the act of strumming guitar strings or the other visual elements shown in Figure 5.2. The musical-cultural signifieds are the various concepts and meanings gained from the musical and visual indicators in the context of learned musical performance event conventions.

Performance signifiers and musical-cultural signifieds can also be explained using traditional acoustic instruments. In watching a musician play a bowed violin, the movements made in playing the instrument are the performance signifiers. The audience's experience is built on their understanding of what they see. These understandings are the musical-cultural signifieds.

As indicated by the formula in Figure 5.3, causation is a critical multiplying factor, as there can be no significant human musical-visual agency without causation. To put it another way: causation is the substance behind the symbols. The performance signifiers and musical-cultural signifieds are the respective symbols and interpretations. They can add or subtract from each other, and may increase or reduce the human musical-visual agency. The model shown in Figure 5.3 also indicates how much human musical-visual agency is a critical but not exclusive part of liveness.

5.1.4 General equipment hierarchy and human musical-visual agency

Before exploring innovative and experimental systems and techniques to augment live musical performance events, it is worth considering what equipment will be used and how much human musical-visual agency this equipment is likely to afford. For a broader description and context of these different types of equipment, see Section 4.3.

The visual pseudo-live elements of Figure 5.2 can potentially give a narrative to machine processes, which is one way to communicate these otherwise obscured operations. Aside from the approach of visual pseudo-live elements, different equipment is likely to afford greater or lesser amounts of human musical-visual agency. Figure 5.4 shows a general equipment hierarchy and the relative degree of human musical-visual agency typically associated with it. The equipment at the top is likely to afford more human musical-visual agency than the equipment at the bottom.

As a general trend, equipment is likely to afford more human musical-visual agency if two criteria are fulfilled: the equipment has one specific use and that use has been widely established. Fulfilling these criteria means that the equipment is likely to be more strongly associated with clear performance signifiers and musical-cultural signifieds (discussed in Section 5.1.3.5).

Take, for example, a comparison between a bowed violin and a laptop keyboard and trackpad. A general awareness of Western music is likely to establish that a long, drawn-out bow movement on a violin produces a characteristic sustained violin tone. In contrast, a laptop keyboard and trackpad are designed to afford a wide range of tasks, mainly clerical, but highly flexible in application. When using a laptop keyboard and trackpad to create music, without highly specialised knowledge and a close-up view of both peripherals and screen, it is almost impossible to observe any clear and specific causal connection to the music being made.

Each type of equipment and its human musical-visual agency is discussed in the following sections.

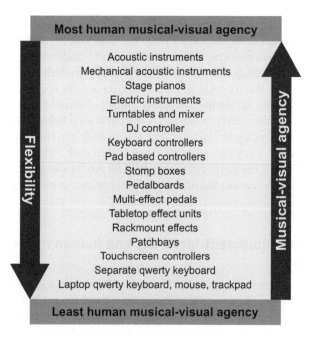

FIGURE 5.4

General equipment hierarchy of human musical-visual agency

5.1.4.1 Acoustic instruments

Human musical-visual agency is exemplified in instruments that have long established a convention of fine expressive control, such as the violin. As discussed in Section i.1.4, an expert violinist has a high degree of fine expressive control. The movements of the violinist clearly and directly cause variations in pitch and dynamics. For example, playing a note with vibrato by moving the hand on the neck back and forth, pivoting on the finger touching the string causes expressive fluctuations in pitch.

5.1.4.2 Mechanical acoustic instruments

The next type of equipment with less human musical-visual agency comprises instruments that exchange the fine expressive control mentioned previously for more polyphony, such as the piano and harpsichord. Because more notes are produced, there is less opportunity for fine expressive control, so the human musical-visual agency is not as clear.

5.1.4.3 Stage pianos

The next type of musical equipment is the stage piano (explained in the context of other keyboards in Section 4.3.3.1). A stage piano is an electronic keyboard with functions typically limited to playing several instruments that tend to be played by piano keyboards (e.g., piano, organ, harpsichord). When a stage piano is played, it is likely to be clear that the sounds created are directly performed using the stage piano.

5.1.4.4 Electric instruments

Electric instruments include the electric guitar, bass guitar, electric piano (e.g., the Fender Rhodes), and tonewheel organ (e.g., the Hammond organ). While played like traditional acoustic instruments, the inherent analogue electronic nature of the tone generation stage means that peripheral sound manipulation equipment is likely to be an essential aspect of playing these instruments – for example, different amplifier characteristics and effects pedals. This additional equipment broadens the sonic potential of these instruments. It makes the tones created dependent on not only the direct manual performance of the instruments, but also the creative use of the peripheral sound manipulation equipment.

It is worth noting that the forward-facing nature of some electric instruments, such as the electric guitar and bass guitar, make the performance signifiers much clearer than instruments played with upward-facing piano-style keyboards.

5.1.4.5 Turntables and mixers

While initially designed for consumer playback, turntables (most notably Technics SC-1200s) were combined with a mixer and repurposed for manual musical curation and as a musical instrument by DJs and Turntablists in the late 20th century. Combining two turntables with a DJ mixer has become the culturally acknowledged musical equipment for playback/beatmatching and scratching. In particular, the human musical-visual agency of moving vinyl discs back and forth on the turntable platter while crossfading on the mixer to make the 'scratching' sound is particularly distinctive.

5.1.4.6 DJ controllers

Modern DJ controllers represent the equipment that allows relatively expensive and fragile vinyl disks to be replaced with a digital audio library. As well as affording a relatively inexpensive and reliable music library, digital audio affords significantly more sound manipulation methods. DJ controllers typically combine two spinning platters with faders, affording the playback/beatmatching and scratching of the traditional turntables and mixer.

DJ controllers are also likely to feature pads, dials, extra faders, and screens to expand the functions of the controller and software. Additional pads, dials, and faders are generic, multi-purpose elements and can be mapped to a wide range of actions, giving a massive amount of flexibility and customisation of workflow. Moving the spinning platters back and forth affords the human musical-visual agency of turntables. However, because of the flexibility and customisation that the additional pads and dials afford to workflow, their use is likely to demonstrate less human musical-visual agency.

5.1.4.7 Keyboard controllers

As discussed in Section 4.3.2.1, keyboard controllers are hardware peripheral devices typically used to control software instruments using piano keys and other input elements such as modulation wheels, faders, and dials. Unlike stage pianos, keyboard controllers are designed to be highly flexible and configurable. Keyboard controllers are likely to have 'synth action' keys (rather than weighted or semi-weighted keys, as discussed in Section 4.3.3.1). Many include the generic and multi-purpose pads, dials, and faders discussed in the DJ controllers section.

As mentioned in the DJ controllers section, these generic and multi-purpose pads, dials, and faders can be mapped to a wide range of actions. However, it is also common practice for the piano keys of a controller keyboard to be mapped to particular actions other than the more traditional triggering of an instrument. These actions may include clip launching or effect triggering, for example. This potential repurposing of the piano keyboard combined with the generic and multi-purpose nature of the pads, dials, and faders means that their use is likely to demonstrate even less human musical-visual agency than DJ controllers.

5.1.4.8 Pad-based controllers

As discussed in Section 4.3.2.1, pad-based controllers are also hardware peripheral devices, typically used to control software instruments using a grid of pads. Unlike keyboard controllers, pads do not have such a strong association with a single instrument (as keyboard controllers do with the piano). The most common pad-based controllers are 4×4 grid pad equipment such as the Akai MPC series and the 8×8 grid pads of Ableton Push and Novation Launchpad.

The Akai MPC series of equipment helped to establish the 4×4 grid as a percussive means of triggering samples, especially drum beats. Accordingly, pad controllers with a 4×4 grid are likely to be associated with drum triggering, particularly for hip hop music. In contrast, 8×8 grids of pads are well suited for launching clips in non-linear clip launching modes and general live set management.

These are broad generalisations, but the flexible and user-configurable nature of modern pad controllers means that they are likely to demonstrate less human musical-visual agency than keyboard controllers.

5.1.4.9 Stomp boxes

A stomp box is a single switch effect pedal, discussed in Section 4.3.4.2.1, with the switch activating and deactivating a particular effect. There is often a performative element in the use of a stomp box. An example is in the music video for the 1997 indie-rock hit 'Song 2' by the band Blur, where a RAT distortion pedal is conspicuously stomped upon by the guitarist as a visual indicator of incoming sonic excitement. Notably, this visual cue summons many sonic changes: the previously band-limited drums transform into hyped rock drums. These are joined by a heavily distorted bass guitar, which – in a then somewhat novel variation to the standard rock trope – is far more distorted than the electric guitar.

The popularity of stomp boxes has spread beyond electric guitar performance, and they are used widely in other genres, for example, by the electronic artist Rival Consoles. The broad range of different effects available as stomp boxes also detracts from the inherent human musical-visual agency, especially when the change made to the sound is not immediately apparent.

5.1.4.10 Pedalboards

Different stomp boxes are often combined to make a pedalboard of multiple effects. While each stomp box will make its specific change to the sound, it is rarely apparent to an observer what these changes are, so this combination of pedals is likely to further reduce the human musical-visual agency.

5.1.4.11 Multi-effect pedals

Multi-effect pedals are typically the size of several stomp boxes with multiple switches configurable by the user and often afford complex combinations of effects to be activated as a part of a performance. As a common trend shown by this hierarchy, this flexibility indicates an accompanying lack of human musical-visual agency.

5.1.4.12 Tabletop effect units

As discussed in Section 4.3.4.2.2, tabletop effect units function in much the same way as multi-effects pedals, except that they are designed to be operated manually (rather than activated by foot). Because of this, tabletop effect units are likely to have many more flexible and configurable controls that are adjustable live. As discussed previously, this indicates a further lack of human musical-visual agency.

5.1.4.13 Rackmount effects

Rackmount effects are discussed in Section 4.3.4.1 and are equipment typically designed for use in the recording studio or as part of live sound setups. Because of this, the controls of rackmount effects are likely to be numerous and complicated, and are often forward-facing (rather than upward-facing), affording very little human musical-visual agency.

5.1.4.14 Patchbays

The routing of rackmount effects is typically managed by using patch cords in patch bays. Patch bays are rows of small sockets typically labelled in small writing. While some electronic artists, such as the Chemical Brothers, have been known to patch equipment as a part of a live set, this is a challenging task which affords hardly any specific human musical-visual agency.

5.1.4.15 Touchscreen controllers

Touchscreen controllers are discussed in Section 4.3.2.3. Partly due to the accessibility of the technology, and to their ease of use and highly visual interface, touchscreens make very effective and extremely flexible instrument controllers. The flipside of this flexibility is that, even with the older dedicated musical devices such as the Jazzmutant Lemur, there are very few controller conventions beyond XY pads and faders. As with the other devices that are so highly flexible, there is a correspondingly low amount of musical-visual agency. There is also the issue that touchscreens are likely to be facing away from the audience, further obscuring any potential performance signifiers.

5.1.4.16 Separate qwerty keyboard

The QWERTY keyboard is discussed in Section 4.3.2.4 as a near-ubiquitous computer peripheral device. Because the QWERTY keyboard has such a strong association with office-based workflows and affords such a broad range of uses, it is not easy to repurpose these devices to demonstrate human musical-visual agency. However, a notable exception can be found in the 2015 setup of Tim Exile's[3] Flow machine, which elegantly repurposes a separate QWERTY keyboard.

5.1.4.17 Laptop QWERTY keyboard, mouse, and trackpad

The equipment at the bottom of Figure 5.4 shows the most generic input peripheral devices: the laptop QWERTY keyboard and mouse or trackpad. It is difficult for these devices to demonstrate any significant human musical-visual agency because of three main factors:

- the broad range of uses afforded by a laptop QWERTY keyboard and mouse
- their strong association with office-based work
- their proximity to the computer, which reflects the familiarity, flexibility, and 'already-there-ness' of this equipment.

5.1.5 Mapping

Mapping is a concept core to any system of chance, reaction, or interaction.

As the name suggests, mapping represents how parameters are altered in a system – specifically, how a source input is 'mapped' or assigned to a destination output. A basic example of mapping is the modulation mappings discussed in Section 3.4 – the mapping of a low frequency oscillator (LFO) to the filter cutoff, which gives movement to the tone in the form of the signature 'wob-wob' sound. The modifier is typically used to allow external adjustment – for example, using a MIDI controller modulation wheel to give depth adjustment in live performance.

In the broader context of systems for electronic music performance, the mapping opportunities are potentially endless. In ever more complex arrangements, many inputs can be mapped to an equally huge range of outputs. Inputs may include gestural or direct controller data, or information derived from the performance or audience. Outputs can be any parameter, either mappable using MIDI Learn or accessed using a more flexible system such as Max for Live or Reaktor. One of the biggest challenges for modern music performance is making these mappings meaningful to the audience, to afford more musical-visual agency (see Section 5.1.3.3).

There are four basic categories of general mapping scenarios shown in Figure 5.5: one-to-one, many-to-one, one-to-many, and many-to-many. The sources are numbered, destinations are lettered, and three dots below the sources and destinations indicate as many more sources or destinations as necessary.

The following are simple examples of the four mapping scenarios to illustrate the concepts, serving as illustrative introductions to each scenario.

5.1.5.1 One-to-one

One-to-one mapping is the simplest type, where a single source is mapped to a single destination. The example given in Section 5.1.5 is a one-to-one mapping where an LFO (the source) is mapped to alter the cutoff of a low pass filter (the destination) with additional depth control from a MIDI controller modulation wheel (the modifier). This kind of machine source mapping is likely to offer reliability and control but not much musical-visual agency. If the source is an external input that can be demonstrated to the audience and the destination is a prominent musical element, then this kind of one-to-one mapping may offer significant musical-visual agency.

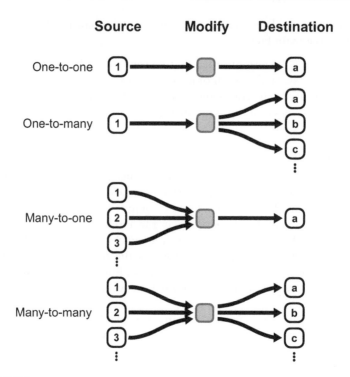

FIGURE 5.5
Four general mapping scenarios

An example of a musical pseudo-live element using one-to-one mapping is the plugin RB.tempo, one of the reactive backing devices I created for Ableton Live, which is explored further in Section 5.5.1.1.

5.1.5.2 One-to-many

One-to-many mapping is where one source is mapped to many destinations – for example, a single MIDI controller (typically a piano keyboard controller) that is set up to trigger multiple sounds. A standard one-to-many mapping may be a monophonic bass synthesiser part that layers many different tones to create a lush and full sound or a single finger chord that triggers multiple evolving arpeggios.

Another one-to-many mapping example is if a single volume source alters the dynamics of several other audio tracks and virtual instruments. This mapping can be done with the plugin RB.dymanics, one of the reactive backing devices I created for Ableton Live, which is explored further in Section 5.5.1.2.

5.1.5.3 Many-to-one

Many-to-one mapping is where many sources are mapped to a single destination – for example, the traditional group performance model of multiple performing musicians (the sources) collectively establishing the ongoing tempo (the destination) of a live musical performance.

An electronic equivalent of this is an ongoing development of the RB.tempo device (discussed in Section 5.5.1.1), which sums and weights different tempo inputs and then alters the tempo. (Currently, multiple instances of RB.tempo can be used, but they alter the tempo individually rather than collectively.) A multiple-input RB.tempo device is likely to be a part of future reactive backing plugins.

5.1.5.4 Many-to-many

Many-to-many mapping is where many sources are mapped to many destinations.

A musician's response to a group musical performance environment can be seen as a complex case of many-to-many mapping. The inputs are perceived changes to tempo, dynamics, groove, and tone/timbre derived from other musicians' combined sound and movement. The outputs are changes to the musicians' tempo, dynamics, groove, and tone/timbre. While in this simplified example, the inputs could be individually matched to outputs as one-to-one mapping, musical performance is likely to be a more holistic combination of these and many other factors.

It is worth noting that effective replication of the human musical response is a clear starting point for mapping scenarios; however, it is not the only goal of mapping a pseudo-live system. A pseudo-live system can be as complex, varied, and flexible as imagination, software, and hardware allow. The mapping may get to a point where it becomes so convoluted that it loses coherent meaning and becomes separate from the intentionality of the performance. At this point, it is worth asking whether pre-prepared music would not be more suitable.

5.2 RECORDING STUDIO AND ELECTRONIC MUSIC TECHNIQUES IN LIVE PERFORMANCE EVENTS

This section discusses sound manipulation techniques that are typically and traditionally restricted to the recording studio. This restriction is partly because of technological and environmental limitations and a degree of path dependency and resistance to change. This section discusses the different types of techniques, and methods of application in live performance events. While these techniques are often creative and corrective, the editing, effect application, or parameter changes are separate from the performance, just as they are in the recording studio's control room. The use of sound manipulation techniques as instruments or performance tools (i.e., playing effects) is explored in Section 5.3.

For the sake of workflow and technique comparison, it is helpful to split recorded music into two categories: 'music created in the recording studio' and 'electronic music'. (This continues the discussion of the same categories made in Section 3.) The defining characteristics of music created in the recording studio, electronic music, and live performance events are outlined here.

Music created in the recording studio: Music created in the recording studio is produced using sources predominantly captured using microphones – for example, acoustic sources such as pianos or vocals or amplified instruments such as electric guitars played through amplifiers. Because these sources create sound, the optimal environment for tracking is a control room for monitoring the recording and one or recording spaces, such as live rooms and booths, separated by multiple walls and doors. These are the spaces that make up a recording studio.

Electronic music: Electronic music is created using predominantly direct sources – that is, electronic instruments or computer plugins. Because these sources typically make analogue or digital audio signals, separating the control room and recording rooms is not as necessary. Accordingly, environments set up for electronic music production are typically one single space. They are often called project studios (or post-production studios) to indicate a variation on a typical recording studio.

The typical production phases that make up the recorded music categories of 'music created in the recording studio' and 'electronic music' are shown in Figures 5.3a and 5.3b. It is worth noting that these categories are very different models of creative practice and, as with many models of creativity, are often more helpful if considered as a spectrum rather than binary states. Some music may be made using only one of these approaches, but a significant amount of music is created by combining them.

Sticking to only one model of creative practice (i.e., electronic music or music created in the recording studio) gives a project more procedurality (procedurality is introduced in Section 4.4.2) – that is, the workflow has a more well-defined and efficient structure but is inherently more rigid. Conversely, combining these models gives less procedurality, allowing more creative freedom, but finding a completed form is likely to be more challenging.

A common problem in electronic music is the difficulty or even impossibility of performing that music because the sound creation techniques are so heavily dependent on processes that do not all occur in real time. While not quite to the same extent, studio music also suffers from a similar problem. The expectation of a live performance event to match the sonic experience of a recording is likely to lead to disappointment.

The term 'live performance events' encompasses the two main production areas of 'live musical performance' and 'live sound'. It is worth noting that the live musical events discussed here require amplification for the audience to hear effectively – that is, they are not traditional orchestras, which traditionally are purely acoustic productions.

'Live musical performance' is the term used here for musicians performing music in an area separate from the audience, typically called a stage (this may include spaces from DJ booths and pub corners to stadium stages). 'Live sound' is the term for the sound manipulation techniques carried out by sound engineers to optimise the sounds of the musicians and play that optimised signal through loudspeaker systems. These live sound techniques typically take place 'front of house' – that is, within the audience, so that the immediate results of the applied techniques can be heard. The typical production phases for live performance events are shown in Figure 5.6 column c.

5.2.1 Production phases

A summary and comparison of the process phases that conventionally take place in the production of electronic music, music created in the recording studio, and live performance events are shown in Figure 5.6. The recording studio processes in 5.3b are shown as the central column, to which electronic music and live performance events are compared because this is the historically established space for producing musical recordings (which are the culturally dominant form).

5.2.1.1 Recording studio phases

Production phases of a conventional workflow in the recording studio are shown in Figure 5.6 column b. It is to be noted that these phases represent a highly generalised model of workflow, effective here in allowing convenient comparison between these three different frameworks of musical/cultural engagement. This model is not intended to demonstrate either detailed or exclusive best practices.

A summary of the main recording studio phases shown in Figure 5.6 column b, is as follows.

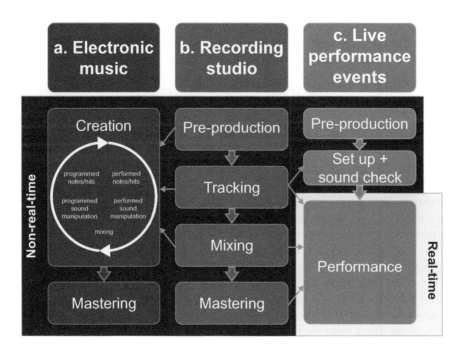

FIGURE 5.6
Comparison of typical production process phases

Pre-production: The pre-production phase covers the planning and preparation of music, musicians, instruments, and studio equipment. While some audio may be prepared in advance in a traditional recording studio workflow, most music is likely to be captured in the tracking phase.

Tracking: The tracking (or recording) phase covers recording musical performances, predominantly using microphones. Also included in this definition of 'tracking' is the process of setting up and testing different microphone, instrument, and environment arrangements, as well as the techniques of overdubbing (recording after an initial recording) and re-amping (sending a recorded signal through an amplified system to capture with a microphone).

Mixing: The mixing phase covers the sonic manipulation and combination of tracks, typically into a two-track 'mix'. (See Chapter 2 for more details about mixing techniques.)

Mastering: The mastering phase covers the manipulation of the mix in preparation for a specific release and format.

5.2.1.2 Electronic music production phases

Unlike the main production phases of the traditional recording studio process, in electronic music production, the pre-production, tracking, and mixing phases are likely to be somewhat combined. This combination is because electronic music has a higher proportion of music that is built rather than recorded (see Section 3 for an explanation of these terms). Also, any recorded performance often has as much to do with making the notes as with the creative application of sound creation techniques. Because of this, these elements of programming, performance, and sound manipulation are all combined in Figure 5.6 column a into a single interwoven creation process.

In a commercial music release workflow for electronic music, the mastering process is likely to remain discrete in the same way as the recording studio process to ensure appropriate sonic coherence across the release and format (e.g., an album for lossless streaming).

5.2.1.3 Live performance events

As shown in Figure 5.6, a live performance event is likely to have a distinct pre-production period as with a traditional recording studio production. Live performance event pre-production is similar to recording studio pre-production outlined in Section 5.3.1.1, except for any pre-prepared performance elements – for example, any combination of MIDI clips, audio clips, samples, or generative devices. These elements and the systems to play them must all be prepared and tested beforehand for consistently successful performances.

As also shown in Figure 5.6, the main production phases of a typical live music event differ from the traditional recording studio phases in two significant ways. The first is with the addition of a sound check phase. In this case, sound check is the setting up of the public address (PA) system and microphones, and the musicians performing to check the equipment and the environment. This setup task is loosely comparable to the microphone setup, and initial testing/ experimentation typically occurs in the earlier stages of tracking in the recording studio audio. The second way that the production of live performance events differs from the traditional recording studio phases is in the performance itself.

After the sound check, the performance will combine the playing of instruments and pre-prepared performance elements. The sound engineers will then mix these elements, and that mix will be amplified and transmitted as sound to an audience through loudspeakers. These processes are loosely comparable to the recording studio's later stages of the tracking phase and the mixing and mastering phases. However, the live sound processes and the musical performance occur in real time. All other phases do not occur in real time, so they are shown as 'non-real time' in Figure 5.6.

5.2.2 Time and space separation

The fact that live components of a live performance event are all processed in real time leads to a core consideration for any attempt to apply recording studio techniques in live performance events. In live performance events, production processes are not significantly separated in either time or space from the musical performance.

5.2.2.1 Separation in the recording studio

Space separation from musical performance in the recording studio: The core spaces of a recording studio are the control room and the recording room(s). The recording and mixing typically occur in the control room, and the studios are typically where the music is performed. The control room is as separated as possible from the sounds played in the studio to give maximum control and the best environment for making critical decisions. Multiple doors, rooms within rooms, and as much physical separation as is practical are used to achieve this separation.

Time separation from musical performance in the recording studio: A defining aspect of music produced in the recording studio is to be able to record multiple instruments on different tracks and to play these tracks back later in time, over and over, to apply other sound manipulation techniques.

This separation in space and time allows for audio sound manipulation techniques that give modern music its characteristic 'larger than life' or 'hyped' sound. Specifically, the techniques that create this hyped sound can be summarised as editing, dynamic processors, and effects (discussed in more detail in the context of the recording studio in Chapter 2). How these techniques can be adapted to live performance events is discussed in Section 5.2.3.

5.2.2.2 Separation in live performance events

Space separation from musical performance in live performance events: As mentioned in Section 5.2, typically, the area for a musical performance is a stage, and the live sound mix takes place front of house. The stage and projected audio are separated only by a small amount of air and the direction of sound projection – nothing to give any significant sound separation compared to a recording studio.

Time separation from musical performance in live performance events: For a performance to be effective, the mixing and transmission to the audience should ideally be as immediate as possible. Latency (delay) times for live sound systems can be as low as 1 millisecond (ms) with minimal processing, but each effect tends to add more latency. Generally, less than 10 ms latency is acceptable (depending on several factors, including 'flight time' of 2.9 ms per metre from the monitor speaker). It is worth noting that live sound systems are likely to have significantly lower latency than DAWs, where the overall latency tends to be more than 10 ms.

Compared to a typical recording studio, which offers as much space or time separation as required to apply editing and effects at leisure, live performance events have virtually no space or time separation from the musical performance. These factors significantly influence how editing, effects, and processors can be applied in live performance events.

5.2.3 Applying recording studio and electronic music techniques in live performance events

In considering the application of recording studio techniques in live performance events, it is worth outlining the conventions of live sound, the recording studio, and electronic music.

5.2.3.1 Live sound conventions

In a traditional live performance event, the musicians on the stage play the music, creating the musical sound for each instrument. The engineer(s) front of house manipulate these musical signals and, as mentioned earlier, optimise these sound signals for effective projection through loudspeaker systems. Typically, live sound engineering is made up of mainly corrective use of dynamic processors and effects, aimed at the audience hearing a more or less authentic version of what is played on stage.

The dynamic processors and effects used in live sound are similar in fundamental operation to those used in the recording studio. However, due to the lack of time and space separation, the cost and the need to keep latency as low as possible, it is likely that fewer dynamic processors and effects will be used in live sound engineering compared to the recording studio.

It is worth noting that continued technological developments in digital live sound equipment give access to significantly more dynamic processors and effects. More dynamic processors afford gating and compression for individual channels, so that sound can be bigger and have more impact. More high-quality reverb and delay effects afford a richer and lusher sound. Together, more dynamic processors and effects let live sound engineers get closer to the corrective techniques of the recording studio, but the lack of time and space separation remains. Because of this, dynamic processors and effects for live sound tend to have controls that allow for quicker and simpler overall operation.

5.2.3.2 Recording studio and electronic music conventions

With few exceptions, sound manipulation techniques of the recording studio and electronic music make extensive use of time and space separation, allowing more control and complexity than the live sound environment.

There is also the convention that sound manipulation techniques can be used creatively as musical tools in the recording studio context. Early examples of highly creative use of recording studio tools include electro-acoustic composers such as Karlheinz Stockhausen and bands such as The Beatles.

As illustrated in Figure 5.6 column a, these sound manipulation techniques are core to electronic music creation. As discussed in Section 3, non-real-time production of electronic music (and, to a lesser extent, music created in the recording studio) lends itself to a workflow that can be more like building or sculpting rather than performance. However, vital parts of this sonic sculpture may also be a studio performance of sound manipulation techniques – for example, the cutoff control of a resonant filter swept expressly. The term 'studio performance' is used here to indicate the performance of electronic music or music created in the recording studio to distinguish it from live performance events. Studio performance can apply to the performance of acoustic and electric instruments and sound manipulation techniques or electronic instruments for which the sound manipulation techniques are an inherent part of the instrument, such as synthesisers. Studio performance is also a part of a non-real-time production process and is a very different practice to live performance. The general objective for a studio performance is to commit the ideal 'studio take' to a recording in a relatively controlled and unresponsive environment. The general aim of a live performance is to play that music to a live responsive audience and ideally achieve an effective exchange of energy.

The three types of sound manipulation techniques of the recording studio – dynamic processors, effects, and editing – are discussed in detail in Chapter 2. The application of sound manipulation techniques of the recording studio and electronic music in live performance events is explored in the following sections.

5.2.3.3 Main differences between techniques

Before exploring the details of applying recording studio and electronic music techniques in live performance events, it is worth establishing a general overview of the differences between standard workflows in these areas.

Corrective recording studio and electronic music techniques tend to be used more deeply and extensively than the equivalent live sound techniques – for example, silencing tools rather than gating and far more deep and clinical use of compressors combined with equalisers (EQs).

As discussed, live sound techniques are predominantly corrective. However, in the recording studio and electronic music environments, extensive creative and corrective sound manipulation techniques are used, including numerous effects to add character. Electronic music production, in particular, is also likely to use many effects that add hype and movement – for example, multiband compression at the instrument level, sidechaining, and the external modulation techniques discussed in Section 4.4.4.

As discussed in the previous sections, applying these techniques in a live performance situation gives many opportunities and challenges, not least due to where these techniques are applied.

5.2.3.4 Location

The location of where these techniques are applied makes a significant difference in how they are used – for example, if they are applied in the front-of-house live sound area or on stage.

As discussed, live sound engineering is most likely to be in the front-of-house live sound area (except for a monitoring engineer, who is expected to be on one side of the stage), and is mainly corrective.

A significant setup consideration for applying recording studio and electronic music techniques in live performance events is the most effective location to apply these techniques. The front-of-house live sound area is the optimum location for overall corrective live sound techniques due to the perspective afforded by the distance and separation.

Additional sound manipulation, such as the deeper corrective and more expansive creative techniques discussed, is likely to be more effective on stage as the systems of sound manipulation can integrate more directly with the systems of sound creation occurring there. Being on stage allows a series of live performance event processes similar to the electronic music production phase of creation shown in Figure 5.6 column a, where the performance, programming, and mixing operations are combined in the creative process.

For live performance events that use sound manipulation techniques on stage, there is a separation between the systems on stage and the systems in the front-of-house live sound area (typically using a live mixing console that is often combined with outboard). In contrast, the typical recording studio and electronic music setups focus on a single consolidated area of sound manipulation. This single setup of the recording studio and electronic music environments allows different corrective and creative techniques to combine and interact more reliably and effectively.

Separated systems tend to be more reliable because if one part fails, there are others to continue. However, separated systems complicate any attempt to apply similar combined and interacting techniques across systems, especially if there is more than one system on stage. One effective setup is if all of the systems on stage can interact but not rely on each other in their application of deep corrective and extensive creative techniques of the recording studio and electronic music environments. The front-of-house live sound area would then manage general balance, overall compression, EQ, and venue-specific sound factors, allowing greater focus on a more mastering workflow and traditional live sound.

Specific instances and techniques are discussed in the following sections. However, the performance must be planned with a detailed awareness of what the different systems are doing, especially between all levels of stage systems and live sound systems.

5.2.3.5 Categorisation and application

Similarly to Chapter 2, sound manipulation tools are categorised into three main types: dynamic processors, effects, and editing. These categories are mainly due to the different workflows inherent in typical use; however, there is a significant crossover between these types, particularly when a creative and intentionally atypical use is applied. A clear awareness of these practices allows for more effective rule-breaking. Without this awareness, a long and laborious process of reinventing the audio wheel is likely.

A review of the typical workflows for each category is as follows:

- Dynamic processors typically reduce, remove, or boost specific signal elements. Standard practice is to insert into either single channels (more common) or aux/group channels (less common).
- Effects typically add signal elements. Standard practice is to send a proportion of the signal on several channels to an aux or send channel (more common), or to insert on a single channel (less common).
- Editing is a typically manual procedure involving changing audio or midi in some way – for example, removing unwanted clip elements by selecting and deleting them.

5.2.3.6 Dynamic processors

Two techniques central in creating the 'larger than life' or 'hyped' sound of modern popular music are dynamic range compression with significant make-up gain and boosting high and low frequencies (or cutting mid frequencies) with equalisation. Compression tends to boost lower-level signal elements across the frequency range, resulting in more energy in the frequencies that EQ can boost. Specifically, there is likely to be more energy in the high and low frequencies, typically associated with exciting or loud-sounding audio.

Along with the maximisers discussed in Section 2, it is hard to overstate the importance of dynamic processors in creating 'hyped' and impactful music. However, extensive use of these techniques is likely to cause howling feedback in a live performance event precisely because of this low-level and wide range of frequency signal elements that are boosted. These boosted signal elements are more likely to be picked up by microphones due to the lack of separation between microphones and loudspeakers. Consequently, this causes a feedback loop as the signal is amplified and picked up over and over until the signal reaches a howling saturation point.

It is clear then why live sound engineers need to be aware of the sound manipulation techniques used on stage, especially in the case of dynamic processors. For example, if compressors and equalisers are stacked up in the systems used on stage (i.e., on a vocal microphone), and the live sound engineer also applies compression and equalisation without awareness of the compressed dynamic range and without careful pre-fade listening, this also increases the risk of howling feedback.

There is no one overall correct way to apply these techniques. However, one starting point is for the on-stage compression to do more transient massaging and only minimal level evening-out, leaving any assertive level evening-out to the live sound engineers as they control the overall balance.

A careful combination of a gate and expander is a useful method to manage the risk posed by raised low levels. The gate can be set to remove any very low-level noise or signal low enough to be unnecessary. The expander can be set to reduce the moderately low-level signals, which can smooth out more aggressive gating and minimise the risk of feedback.

A simplified illustration of the transfer curve caused by a gate, expander, compressor, and limiter is shown in Figure 5.7a, which plots the incoming signal amplitude on the x (horizontal) axis against the output amplitude on the y (vertical) axis. The faint grey 45-degree diagonal line shows a system with no change, a bypassed input straight to output for comparison. Figure 5.7b shows the same transfer curve with +12 decibels (dB) of make-up gain.

The dotted lines show the result of the different dynamic processors, explained in the following sections in terms of amplitude.

Figure 5.8 shows a screenshot of plugins and settings that create the dynamic manipulation shown in Figure 5.7b. All plugins are part of Ableton Live except for the Apple AUDynamicsProcessor, as this allows downward expansion (at the time of writing, a free windows alternative was the Floorfish Virtual Studio Technology (VST) plugin by www.digitalfishphones.com).

Gate: Line i, in Figure 5.7a, shows how a gate with an infinite floor and a threshold of -42dB responds. Below -42 dB, no signal is passed. As the signal rises over -42 dB, the gate opens, and the signal is let through.

Expander: The expander shown in Figure 5.8 is a downward expander (the most common type). In a downward expander, the dynamic range is expanded downward from the threshold, so that quiet signal elements get quieter. In the example of Figure 5.7a, line ii, the threshold is set at -36 dB. The ratio sets the gain reduction

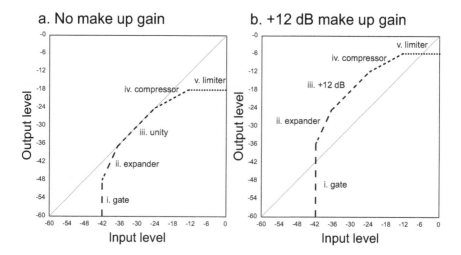

FIGURE 5.7

Transfer curves caused by the combination of a gate, expander, compressor, and limiter below the threshold. In this case, the ratio is 2:1 – that is, for every one dB under the threshold, the signal is reduced by 2 dB. In Figure 5.7a, line ii, where the input goes from -42 dB to -36 dB (a range of 6 dB), the output is -48 dB to -36 dB (a range of 12 dB). Put another way, the input level of -42 dB is output at a lower -48 dB, input -39 is output at -42 dB, and -36 dB is output at -36 dB (the threshold).

Unity: In the example in Figure 5.7a, line iii, the signal is unchanged from -36 dB to -24 dB. In applying these techniques to replicate typical recording studio practice, this signal portion will be boosted by the make-up gains of any compressors or limiters. Because of this, the example of a signal raised by 12 dB overall is shown in Figure 5.7b, and line iii is labelled +12 dB, parallel to the grey bypass line.

Compressor: The compressor segments in Figures 5.7a, and b, show a compressor reducing the dynamic range of a signal over a threshold so that loud signal elements get quieter. The amount of reduction above the threshold is set by the ratio, shown in lines iv. In Figure 5.7a, the threshold is set at -24 dB, and the ratio is 2:1 so that for every 2 dB over the threshold, only 1 dB is output. Accordingly, as the input signal goes from -24 dB to -12 dB, the output is half that: -24 dB to -18 dB.

Limiter: A limiter is essentially a compressor with an infinite ratio of gain reduction. So when the signal reaches the threshold, the output is limited and will not go over this setting. Depending on the attack, lookahead and design characteristics, many limiters may briefly exceed the threshold. (To guard against this, especially in mastering, a 'true peak limiting' mode is sometimes offered, ensuring that no peaks will go over the limiter threshold, but this is not likely to be necessary in a mix.) In Figure 5.7a, the limiter's threshold is -18 dB. Due to the effect of the compressor before the limiter, this corresponds to an initial input level of -12 dB. The gain is set to 12 dB of signal boost to finally bring the signal up to the levels shown in Figure 5.7b. (It is worth noting that this is a particularly aggressive boost, it is shown here to demonstrate the point clearly.)

This chain of four dynamic processors is quite a simple one. Still, it can be clearly seen in Figure 5.7b that a significant part of the medium to loud levels of the signal are +12 dB louder: more than four times the acoustic power, which is likely to sound a little over twice as loud. In an environment with a PA, this needs careful management. Compared to traditional acoustic and electrical instruments, signals processed through a computer on stage can have this kind of

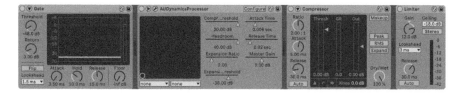

FIGURE 5.8

Plugins and settings that create the dynamic manipulation shown in Figure 5.7b (Ableton Live: Gate, Compressor, and Limiter – Apple: AU Dynamics Processor)

deeper dynamic processing, giving a more hyped and impactful sound. However, it is likely to need rigorous venue-by-venue testing to ensure howling feedback does not occur.

This chain of manipulation can, of course, be applied in the live sound area. If no integration with other systems on stage is needed (e.g., for either chance or reactive systems), this may be more desirable depending on the setup's needs.

Absence of silencing tools: If, however, the intention is to apply combined, interactive, deep corrective, and extensive creative techniques of the recording studio and electronic music environment, then, as mentioned in Section 5.3.3.4, these systems are likely to be located on stage. The most significantly absent tool is the one afforded by space and time separation: the editing operations offered by silencing tools such as strip silence, discussed in Section 3.3.1.2.

Without the aid of time travel, it is not possible to apply silencing tools to live signals, but more advanced applications of gates can be used. For example, some gates allow intelligent transient detection, making them act more like silencing tools. The most notable of these is the Oxford Drum Gate by Sonnox, which, in well-managed circumstances, can deliver results for close-miced drums that come close to those of silencing tools.

Application of dynamic processors: As explained in more detail in Section 3, typical recording studio use of dynamic processors, such as gates, compressors, and equalisers, is to insert them into individual and group tracks. Because a processor inserted into one track with only one instrument will be manipulating only that signal, the amplitude envelope and spectral characteristics are likely to be more straightforward and more consistent. In turn, this lends itself to a single-specific-signal application of dynamic processors.

A processor inserted into a group track is likely to be manipulating a far more complex combination of signals, lending itself to a broader and potentially more apparent treatment.

Having these dynamic processors running as part of a stage setup opens up many possibilities for sound manipulation based on, and to suit, a changing performance. For example, key parameters can be mapped to macro controls and automated so that songs intended to be more intense can have a tightly controlled dynamic range and 'big' sound. In contrast, other songs can be allowed more dynamic range, which affords a more expressive and nuanced performance.

Another plugin type worth considering is fader-riding plugins. In particular, the Vocal Rider by Waves, mentioned in Section 3.1.5, has a live setting with spill rejection and uses a side chain to monitor the instrument mix and adjust up or down accordingly. While this can act a little like a compressor, it is more of a pure-level evening tool and is worth treating with care to avoid too much gain.

Performance data: There is also the possibility of making these parameters reactive to performance data. For example, a more consistently loud drum performance could drive a more compressed vocal sound using the RB.dynamics plugin to alter a macro compression control. (Reactive elements are discussed further in Section 5.5.)

The controls that are automated or altered by performance data may be varying multiple inserts on individual tracks for separate manipulation, which, if used effectively, is likely to maintain clarity and separation in the mix. Alternatively, or as well, these controls may be varying inserts of a group track (or several group tracks) for combined manipulation, which, if used effectively, is an approach that is more cohesive and is likely to suit more dramatic and broad creative alterations to the mix.

It is worth reiterating that these techniques must be handled with great care and managed alongside the front-of-house sound engineer. As mentioned previously, too much gain and compression are likely to create howling feedback. A series of techniques to mitigate the risk of howling feedback is explored in Section 5.3.4.

As discussed in Section 3, the standard recording studio best practice for effects is to create an auxiliary channel per effect, send a proportion of each required signal to that effect, and return the wet signal to the mix.

In a standard live sound setup, the sends and returns would all be managed in the front-of-house live sound area, which can be done if enough channels are sent separately from each on-stage system, which may need to include any groups or buses.

The opposite approach would be to have each system on stage have its sends applied independently – one well-suited to control alteration by automation or performance data, for example. However, in this situation, there is a risk of inconsistent effect application compared to front-of-house controlling all sends.

A combination of the two approaches may be most effective, with any particularly creative effects added on stage and enough stems being sent to the front-of-house live sound area to add more general corrective/consistency reverb. As before, careful systems planning and testing are likely to be necessary for developing effective and innovative setups.

Application of effects: Compared to processors inserted into existing tracks, the additional send tracks that effects typically bring allow a useful separation. Either effects parameters (e.g., depth and time) or the overall level of the send track can be automated or altered by performance data, and the wet or dryness of a mix can also be dynamically varied. Also, lush reverberation can be swapped for swirling modulation effects as required, for example, or all effects can be pulled back for a more dry and more intimate sound.

The downside of this approach is that separate send tracks bring a potential risk of too much signal level as both tracks can be pushed higher, summing to a hotter signal, which may lead to howling feedback. Also, if too much wet signal is added, the mix is likely to lack clarity as the additional elements may obscure the original.

It is worth mentioning that there is a significant overlap between the workflow of processors and effects presented here. Any processor inserted onto a track with a wet/dry control acts like an effect without a separate return fader, and effects can be inserted onto tracks to be used individually. A notable sound manipulation technique that may be interchangeably used as either an inserted processor or a sent effect is distortion, the use of which is dependent on context and workflow.

5.2.3.8 Editing

Editing techniques originate from the recording studio and have been traditionally corrective and manual adjustments applied to either individual audio or MIDI elements (e.g., trimming or fading), or multiple audio or MIDI elements (e.g., silence and batch fades). As such, they are almost exclusively non-real-time sound manipulation techniques. However, as noted in Section 5.2.3.6, a similar outcome to that achieved with silencing tools can also be achieved with gating.

More modern creative editing applications tend to come from electronic music and are core to genres such as intelligent dance music and glitch-core. Again, these are primarily non-real time sound manipulation techniques, although several plugins buffer (i.e., store) a short section of audio and afford a mechanistic type of audio editing. Examples of buffer editing plugins include Beat Repeat and Buffer Shuffler 2. Beat repeat is an Ableton Live plugin for highly rhythmic delays. Buffer Shuffler 2 is a Max for live plugin, which divides a buffered section of audio (i.e., a bar long) into a series of steps that it shuffles and manipulates in a wide range of ways.

A screenshot of Buffer Shuffler 2 is shown in Figure 5.9. The plugin will buffer the signal for a bar and then play back as per the settings. In the example of Figure 5.9, the first four eighth note steps will play back as before, then the second step will be repeated, the seventh and sixth will be played in reverse, and the bar will finish with the fifth step.

Buffer editing plugins can be suitable for creating experimental, often glitch-like sounds. However, when used in a live performance event, they have the drawback of taking some time for the buffer to fill, and the results cannot be adjusted for context as easily as they can in non-real time environments. Of particular use are the randomisation features that buffer editing plugins offer, such as the dice button at the bottom on Buffer Shuffler 2, which, depending on the settings, adjusts the settings automatically.

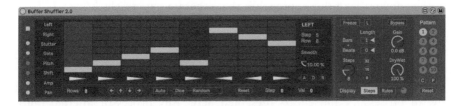

FIGURE 5.9

Ableton Live Buffer Shuffler 2 Max for Live plugin

An editing plugin well suited to a one-shot kind of performative approach is Stutter Edit 2, which gives potentially dramatic results. Stutter Edit 2 is discussed further in Section 5.3.

5.2.4 Reducing the risk of howling feedback

As pointed out, deeper, more creative, and more expansive sound manipulation significantly increases the risk of howling feedback, especially when using compression and equalisation. To most effectively and consistently manage the risk of howling feedback, all signal chain elements should be considered in a system likely to use significant sound manipulation.

Howling feedback is likely to be caused by signals captured using microphones susceptible to picking up their own amplified output from loudspeakers, causing howling feedback as the PA reaches saturation. Unless there is any signal feedback in the chain of a direct sound source such as a synthesiser or sampler, these signals will not cause howling feedback.

Two main factors affect the risk of howling feedback: microphone technique and mixing.

5.2.4.1 Microphone technique

As discussed in Section 2.2.2.1, a microphone changes any acoustic sound energy into electrical energy. Selecting the most suitable microphones and setting them up in the most effective arrangement is key for capturing optimal source sound quality and reducing the amount of unwanted sound the microphone will pick up. Unwanted sound or 'spill' from other instruments is not likely to cause howling feedback. Still, too much spill will reduce the clarity of the overall mix, especially if it is accentuated by heavy compression and equalisation. The amplified sound from monitor speakers or the front-of-house speakers is likely to be the leading cause of howling feedback.

The most common type of microphone used for live performance events is the dynamic microphone because dynamic microphones are less sensitive to signals at a distance (i.e., more than 20 centimetres) from the microphone and tend to be more rugged than condenser microphones. Depending on the overall setup, a dynamic microphone with a unidirectional (i.e., picking up sound mainly from one direction), cardioid, super cardioid, or hyper-cardioid polar pattern is likely to give the best feedback rejection. See Section 2.2.2.1 for a discussion about microphone polar patterns.

As well as this consideration, there are two methods to increase the space separation in a live performance event.

Space separation 1: in-ear monitor headphones: One of the most effective methods of reducing the amount of monitor speaker sound picked up by a microphone is to replace on-stage monitor speakers with in-ear monitors, which give almost complete space separation compared to on-stage monitor speakers. In-ear monitors do not, however, offer any more space separation from the front-of-house speakers.

Space separation 2: baffles/shields/screens: A less common approach is placing something between the unwanted amplified sound and the microphone. For drum kits played at live performance events, clear plastic shields encasing the drum kit or individual baffles mounted on microphone stands can reduce loudspeaker sound and spill. For other instruments, more general-purpose transparent screens – typically folding sheets of plastic – can be placed in front of the microphone(s).

5.2.4.2 Mixing

As explained in Section 5.2.3.6, substantial compression and equalisation are likely to increase the gain of lower-level signals. In a live performance event, lower signals captured live using microphones are likely to include some amplified output from the monitor or front-of-house loudspeakers. Without heavy compression and equalisation, these lower levels would not be expected to cause much of a problem. However, when these signals are boosted, any minor microphone issue will increase the risk of howling feedback.

A widespread example is if a singer cups the back of a dynamic cardioid microphone like the Shure SM58. The SM58 and many others like it achieve their cardioid polar pattern by using a Uniphase design, causing signals entering through the back of the microphone to cancel themselves out. If the cupping of a hand blocks the back of the microphone, for example, the polar pattern is likely to shift to a more omnipolar pattern and pick up more sound behind the microphone. In this case, howling feedback is far more likely, especially if compression and equalisation boost the lower-level signals from the monitor or front-of-house loudspeakers.

If these factors are considered, systems can be tested to establish and manage practical limits. It may be that a more extensive and more reactive range of compression is available using a particular kind of microphone and in-ear monitors.

5.2.5 Pre-prepared performance elements and live musical elements

Pre-prepared non-live or pseudo-live elements for performance events are likely to be produced in the separation of a recording studio and electronic music environment, so the respective sound manipulation techniques are likely to be employed. There is an understandable temptation to aggressively 'hype' these elements, making them particularly impactful and 'big' sounding; however, this is likely to lead to these elements having a noticeably different sound from the elements played live.

Of course, a noticeably different sound may be the desired result and part of the performance aesthetic. Alternatively, if the desired result is to create a balanced range of sounds that combine coherently, it is worth knowing how many non-live, pseudo-live, and live elements are being used, as well as how to incorporate them effectively.

5.2.5.1 Combining pre-prepared performance elements and live musical elements

Depending on how automated or reactive the processors and effects used in a performance event are, it may be possible to use the same or similar processors, and effects in real-time on both the pre-prepared performance elements and live musical elements. Using the same processors and effects is likely to reduce the difference in sound manipulation techniques and give more sonic coherence.

The downside to this approach is that processing pre-prepared elements live will add more processor load to a system. This load can be mitigated by choosing 'core' techniques which are not likely to change and bouncing these beforehand. This evaluation of techniques applied before and during a performance is a balance between system stability and simplicity on the one hand, and parameter variability and consistency on the other.

5.3 PLAYING EFFECTS (EFFECTS AS PERFORMANCE TOOLS)

Deep corrective and extensive creative use of effects is discussed in the previous section in the context of techniques of the recording studio and electronic music environments. ('Effects' is used as an umbrella term for both effects and processors from this point onwards.) Standard practice for corrective effects is to set the parameters of the effect and either leave them fixed or automate carefully so that the effect is appropriately balanced. For creative effects, parameters are often altered using either automation or modulation so that the effect is suitably dramatic. As discussed in Section 5.2, there is rarely any causal connection to these effect changes.

An extension of the creative application of effects is to adjust critical parameters in a musically meaningful manner during performance – that is, 'playing' the effect. The effect parameters that are changed may be a vital musical element of playing the instrument – for example, using a wah pedal rocking back and forth for rhythmic funk guitar parts.

The bypass state of each effect – that is, its activation or deactivation – is the most basic parameter adjustment using a simple on/off toggle operation. Other parameters are likely to cover a gradient of adjustment – for example, depth, wet/dry level, frequency, and rate.

Modern digital equipment gives a vast range of possibilities for using effects as performance tools, often acting like instruments in their own right. The ways that effect parameters can be adjusted as a part of a performance can be split into two categories: direct parameter adjustment and indirect parameter adjustment.

5.3.1 Direct parameter adjustment

Direct parameter adjustment occurs when the musician adjusts a physical control that alters one or more of an effect's parameters in real time during the performance.

Direct parameter adjustment can be applied to either individual or multiple effect parameters. It can use analogue controls, such as a knob on the end of a potentiometer in an outboard hardware unit, or digital controls, such as a knob connected to an encoder in a MIDI controller.

In standard use of outboard hardware units, analogue controls are likely to be hardwired to an individual effect parameter – for example, the feedback control on a delay effect or the frequency control that is swept by a wah pedal. Notable exceptions to this general rule include modular synthesis units such as eurorack modules where control voltage signals can be patched to adjust a wide range of parameters.

For digital controller devices, control elements are easily mapped to one or more effect parameters using DAW or plugin host software.

Direct parameter adjustment of an individual effect parameter has a direct causality, similar to the direct causality of acoustic musical instruments introduced in Section 5.1.3.2. Unlike the context of musical instruments, as discussed in Section 5.1.3.5, this direct causality translates to a different type of human musical-visual agency due mainly to the vast range and flexibility of mapping afforded by many effects, especially digital effects.

As described in Section 5.1.3, human musical-visual agency is determined by the combination of causality, performance signifiers, and musical-cultural signifieds (i.e., how much awareness there is of the musical context).

For more discrete standalone effects designed for performance, such as the wah pedal, one visually apparent control directly adjusts a parameter (i.e., centre frequency). This direct causality combines a distinct performance signifier and a musical-cultural signified, well-established by famous guitarists such as Jimi Hendrix, to create significant human musical-visual agency.

However, as reflected in the general equipment hierarchy and human musical-visual agency in Section 5.1.4, there are many effect parameters and controls. For more generic devices or controllers, as much as there may be direct causality, there is likely to be little human musical-visual agency if the visual performance indicators are unclear and the musical context conventions are not established.

Performance signifiers are likely to include the outward elbow indicating a knob turn, a smooth back/forward movement indicating a fader pull/push, or the even more anonymous finger track on an XY pad. Musical context conventions for playing effects are particularly limited. Beyond the expressive effects for the electric guitar, such as the wah and whammy pedals, the most notable standard for electronic music is the knob turn on a DJ mixer. The knob turn is likely to indicate a filter sweep creating a euphoric build or more tonal fade commonly associated with techno and house electronic dance music genres. This musical-cultural signified is established because the DJ setup is typically a single DJ manipulating complete musical mixes so that attention is already focused and the sound manipulation is likely to be more evident.

As observed in Section 5.1, this trend is clear: the massive power and flexibility of modern setups affording the playing of effects make conveying authentic human musical-visual agency incredibly challenging.

5.3.2 Indirect parameter alteration

Indirect parameter alteration is when a system alters the parameter over time. Typically, the musician will launch either the effect or the alteration, but the alteration is likely to be managed by a modulation source such as an LFO, envelope, or complicated step sequencer.

A suitable example of an effect plugin designed around indirect parameter alteration is Stutter Edit 2, explored further in Section 5.3.3.1. The effects in stutter edit are altered over time using envelopes tied to the DAW timeline, so no effects will be applied if the transport is not running, unlike most other effects, which apply the effect to any incoming signal. As well as Stutter Edit 2, several other plugins designed to be 'played' are discussed in the following section.

5.3.3 Plugin effects as performance tools

Modern plugin effects are particularly well suited for use as performance tools because of their accessibility and flexibility. Accessibility comes from software effects tending to be more cost-effective than hardware effects, especially if a DAW is already owned. Flexibility comes from the ease at which parameters can be combined into macro controls, mapped onto control devices or, for some DAWs such as Ableton Live and Reaper, controlled using the application program interface (API).

Any effect can be played if applied with sufficient depth to make its adjustment apparent, which opens up a massive array of avenues for experimentation. However, as discussed in Section 5.1.1, it is worth checking in with the aesthetic boundaries of the project to ensure that the creative process continues to serve the project's objectives.

Effect bypass: The most basic means of parameter adjustment is effect bypass. In Ableton Live, this can be achieved by clicking on or mapping the plugin activator switch in the top left of the plugin UI. In multi-effect plugins, individual effect bypass can often be managed internally (as well as a global effect bypass).

A small subset of multi-effect plugins are designed specifically for the playing of effects, which are referred to here as 'effect player plugins'. Effect player plugins are designed to trigger and alter effects and critical parameters in a performative fashion. Effect triggering is typically achieved using MIDI notes to activate and deactivate effects. Three notable examples of effect player plugins are The Finger, Artillery 2, and Stutter Edit 2, discussed in Section 5.3.3.1.

Parameter alteration: Effect parameter alteration is typically achieved using MIDI continuous controller (CC) mapping. Alteration can occur either at the DAW level – for example, using Ableton Live MIDI map mode, or at the plugin level – for example, using MIDI learn inside a Native Instruments Reaktor ensemble such as The Finger. CC mappings can either be direct one-to-one mappings or more complex macro mappings where a single control can control multiple parameters.

Examples of DAW level one-to-many mapping (see Figure 5.5) in Ableton Live include effect rack Macros for parameters in the effects racks, and the Max for Live audio plugin MultiMap for parameters across the Live set. A screenshot of MultiMap is shown in Figure 5.10, showing how the input control (which itself can be mapped to a MIDI control) is mapped to adjust the Decay of the DS Kick plugin instrument, the Drive of the Saturator effect, and the Filter drive in Operator on another track. The mapped controls have their dial amount changed from blue to white, as can be seen in Figure 5.10 for the controls on that track.

Even more complicated custom MIDI CC mappings can be created using Max for Live patches that access the Live API. For example, the free RB.CCmodfilter MIDI utility plugin shown in Figure 5.11. RB.CCmodfilter is a MIDI plugin that takes up to 16 MIDI CC messages (selected in the number boxes above the T buttons); it allows value rescaling and offsetting, and using the 'T' button (for MIDI Teach), can map to the AIP driver. The MIDI output of RB.CCmodfilter also filters out any MIDI notes outside the range set on the right of the plugin.

Examples of CC learn mappings at the plugin level include the mappings shown in The Finger in Figure 5.12 in the following section.

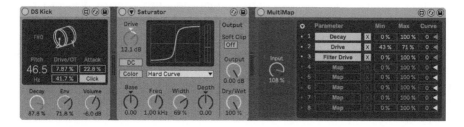

FIGURE 5.10

Ableton Live MultiMap Max for Live plugin and the Ableton plugins Live DS Kick and Saturator

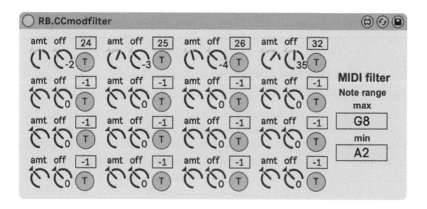

FIGURE 5.11

RB.CCmodfilter Max for Live plugin

5.3.3.1 Effect player plugins

All three effect player plugins discussed here are designed to trigger effects by incoming MIDI 1.0 messages played on a standard MIDI piano keyboard (at the time of writing). Artillery 2 was released in 2007, The Finger in 2009, and Stutter edit 2 in 2020. At the time of writing, all software was well supported.

The Finger: The Finger is a plugin created by the electronic music performer Tim Exile and Native Instruments, which runs as a Reaktor ensemble. A screenshot of The Finger can be seen in Figure 5.12. Effects are assigned individually to each MIDI key, which, when played, applies the assigned effect. Up to six effects can be triggered dynamically – that is, they are routed in the order they are triggered. The corresponding effect for each note can be seen at the bottom of Figure 5.4.2. For example, note E1 is set to trigger the FM 180 effect.

Effect parameters are initially set by the two main controls (labelled 'frequency' and 'amount' in Figure 5.12). These parameters can be directly adjusted by note velocity and modulation wheel CC messages and indirectly altered by the envelope settings.

As shown on the left-hand side of Figure 5.12, controls can be mapped using MIDI learn. However, if more than one effect is triggered, then that control adjustment will apply to all effects triggered.

Artillery 2: Artillery 2 is a plugin created by Sugar Bytes, a screenshot of which can be seen in Figure 5.13. As with The Finger, effects are assigned individually to each MIDI key, which, when played, applies the assigned effect. As can be seen, there are more effect parameters and modulation parameters than The Finger, and most of these parameters can be assigned using MIDI learn. However, due to the number of parameters and their relatively discrete nature, they do not integrate as effectively with modulation (CC1) and velocity.

Stutter Edit 2: Stutter Edit 2 is a plugin created by the electronic music producer BT and iZotope. A screenshot of Stutter Edit 2 can be seen in Figure 5.14. Rather than assigning just one effect to a MIDI key, Stutter Edit 2 assigns a 'gesture' to each MIDI key. Each gesture can have up to 13 effect modules assigned to it,

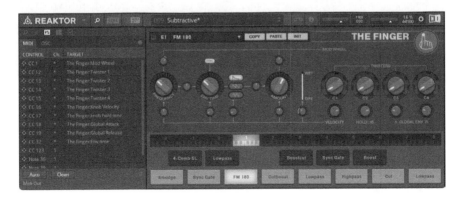

FIGURE 5.12
Native Instruments The Finger plugin

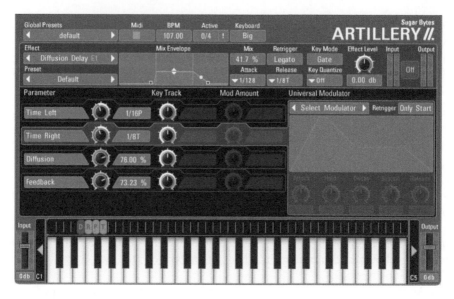

FIGURE 5.13

Sugar Bytes Artillery 2 plugin

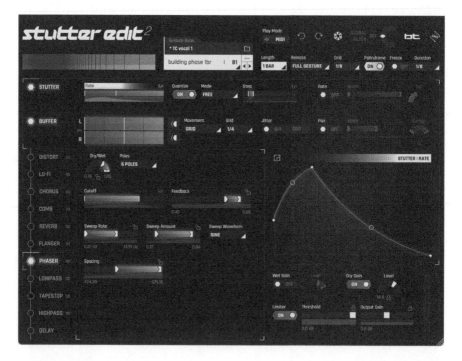

FIGURE 5.14

iZotope Stutter Edit 2 plugin

and each effect parameter is designed to be modulated using its customisable envelopes.

As the name of the plugin suggests, the main effect is the stutter module, which rhythmically repeats either monitored or clip audio depending on the stutter rate. The modulation of the stutter rate is shown in the envelope to the right of Figure 5.14. The possible combination of the 13 highly modulate-able sound manipulation modules allows particularly deep and rhythmic effects. However, apart from the ability to trigger gestures, Stutter Edit 2 allows no direct real-time parameter adjustment. Consequently, liveness is limited to the triggering of the gesture and the machine liveness of the gesture interacting with the buffered audio.

A more limited alternative to Stutter Edit 2 is to use the Ableton Live plugin Beat Repeat and to map the grid parameter to any MIDI continuous physical controller, such as a mod wheel, dial, or expression pedal. In contrast to Stutter Edit 2, this mapping of Beat Repeat does afford direct parameter adjustment.

5.3.3.2 Performance signifiers and musical-cultural signifieds for effect player plugins

As discussed, a piano keyboard is the default controller for effect player plugins for price and accessibility reasons. Accordingly, player plugins are likely to be played using a standard synth-action piano keyboard controller. The drawbacks of a piano keyboard controller for expressive instrument performance are discussed in Section 4.3.2.1. This issue is similar but compounded when a piano keyboard controller is used to play effects. There are three main problems:

- There is no CC capability in the keys (with the rare exception of aftertouch).
- Due to the top-down playing angle and mechanics of the piano keyboard, the performance signifiers are limited.
- An abundance of pre-existing musical-cultural signifieds for the piano obscure attempts to repurpose and establish new musical-cultural signifieds.

Solutions to these problems are likely to include three characteristics:

- expressive controllers (e.g., the controllers discussed in Section 4.3.2.5)
- controllers that the audience (i.e., front facing, can clearly view), and
- multiple high-profile musicians who establish new musical-cultural signifiers.

A specific example of how this can be implemented is shown in Figure 5.15: a QuNeo pad mounted on an Epiphone 335 guitar, which I use in performance for live audio manipulation. Each QuNeo pad is assigned different effects. The first column from the right is assigned to effects in an instance of The Finger. The second column from the right is assigned to effects in an instance of Artillery 2. The third column from the right is assigned to different gestures in Stutter Edit 2, and the final two labelled pads are assigned to another instance of The Finger. The pads fit under the four fingers of the right hand, and key per-pad parameters are mapped and managed using the RB.CCmodfilter Max for Live MIDI effect shown in Figure 5.11. The long modulation strip fits under the thumb for global parameter adjustment.

FIGURE 5.15
Photograph of my guitar-mounted QuNeo setup for live manipulation

5.4 LAUNCHING

One of the key practical differences between playing an acoustic instrument and performing electronic music is that in performing electronic music, musical elements are often launched rather than played.

When an acoustic instrument is played, there is likely to be a significant amount of direct causation. For example, a musician playing the saxophone is the primary cause of continuous musical pitch, duration, dynamics, and timbre. (Direct and indirect causation are introduced in Section 5.1.3.)

Launching (or triggering) is the starting of a machine process and, in its most basic use, is left to run, demonstrating indirect causation. If we take the example of a musician launching a sample using a sampler plugin in a DAW, the primary cause of the continuous musical pitch, duration, dynamics, and timbre is the preparation of the sample and the machine processes of the sampler and DAW – unless, of course, the DAW or sampler controls are modified in real time.

Playing and launching need not always be clearly and distinctly different. For example, an expressive MIDI controller may launch a sample, which may be modified expressively, demonstrating direct causation by continuously varying musical pitch, duration, dynamics, and timbre. This combination of musician and DAW is playing the music.

Alternatively, a key on a piano may be pressed and left to ring out by depressing the sustain pedal, so only the piano mechanics are left to determine the continuous musical pitch, duration, dynamics, and timbre, which in turn is demonstrating indirect causation.

One of the most notable benefits of electronic systems that allow clip and sample launching (i.e., DAWs) is that a vast number of clips and samples can be launched with great simplicity and in ways that can be as intricate and expressive as the musician wishes.

5.4.1 Sample launching

A sample is the most straightforward musical element that can be launched. A sample is a short audio recording that is launched by a sampler. (A more detailed discussion of samples and samplers is set out in Section 3.7b.)

The simplest setup for sample launching is a single standalone sampler with keys or pads, such as the keyboard workstations and pad devices discussed in Section 4.3.3. The sample is instantaneously launched in standard operation by pressing a key or pad.

A single sample launching device is likely to be the most simple and reliable setup; however, this will tend to offer the most restrictive and procedural workflow (the concept of procedurality is introduced in Section 4.4.2). The pros and cons of standalone devices and computer-based systems are discussed in Section 4.3.

5.4.2 Basic non-linear clip and scene launching

As described in Section 3.3.1, a clip is a type of virtual container for musical material, either MIDI or audio information, that is used by DAWs to manage musical material. As described in Section 4.4.2, the traditional mode of operation for DAWs is to launch these clips in a linear timeline fashion; when the transport plays, the clips on the tracks are launched one after each other along the timeline as long as the transport keeps playing. The alternate mode of operation offered by some DAWs is non-linear clip launching. Non-linear clip launching allows the clips to be launched in whichever order the musician wants at the time of performance. Scenes are the rows of clips on adjacent tracks. As tracks are vertical in the session view of Ableton Live, the scenes are horizontal rows. At the time of writing, the DAWs that afford the non-linear launching of clips include Live, Bitwig, Logic, and FL Studio.

The most basic method of launching, apart from clicking on the launch button of each clip, is to map either the individual clip or the scene to a MIDI message or a key which, when pressed, will launch the clip.

The settings of the DAW manage the synchronisation of clips. In Ableton Live, this is set by the global quantise setting, which has a default of 1 bar, so if a clip is launched on the second beat of a bar, the clip will wait to launch until the first beat of the next bar.

Figure 5.16 shows a screenshot of Ableton Live in mapping mode to show that scene 2 (shown in the Master track) is mapped to channel 1 note D3 and the clip in slot 1 of the track '3 Synth Pad' is mapped to channel 1 note C3. When a D3 note on message in channel 1 is received, then, on the next 'one' beat, all clips in the second row (scene 2) will be launched and will repeat. If a C1 note

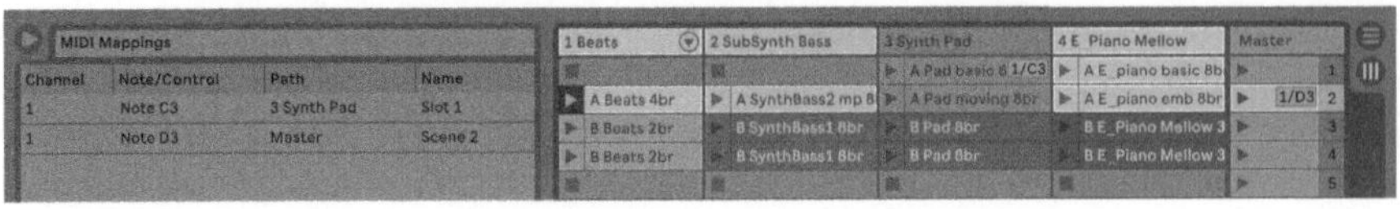

FIGURE 5.16
Ableton Live clips in mapping mode

on message in channel 1 is then received, then on the next 'one' beat, the clip in track '3 Synth Pad', slot 2 will be stopped, and the clip in slot 1 will be launched. After playing through, the clip will repeat, stop, or change based on the follow actions settings. Follow actions allow different clips in the track to be launched based on chance settings and are discussed further in Section 5.6.3.

5.4.3 Moving from a non-real-time production environment to a real-time environment

The standard non-real-time (i.e., not live) production environment for recorded music is the linear timeline of a DAW. Turning a non-real-time project into something that can be reliably and effectively played live (i.e., in real time) is likely to take some optimisation and conversion. If the features of a non-linear clip launching environment are needed, then making this process work effectively is likely to include steps described below in Sections 5.4.3.1 to 5.4.3.4.

If a backing track (two or multi-channel) is required (discussed in Section 3.5.1), this is likely to be a process of just optimising the setup, as outlined in Sections 5.4.3.1 to 5.4.3.3.

The non-real-time sound manipulation techniques involved in producing a recording are typically far less demanding on computer processing than the real-time sound manipulation techniques used in live performance events. The reason for this is that when producing a recording, there are no live signals to process in real time, so the buffer size and corresponding latency time can be much higher. When producing a recording and processing power becomes scarce, tracks can easily be frozen (an automatic 'in track' render that deactivates that track's plugins) to free up processing power.

Also, the stakes of processor overload are far higher in live performance events. A non-real-time crash when mixing is likely to be no worse than an embarrassing inconvenience; however, this is disastrous in a live performance event. A crash is likely to stop the performance if there are no real-time backups. Any pops or clicks (typically due to processor overload or disk access) during a non-real-time mix are simply an indication to freeze tracks or increase the buffer size. In a live performance, pops or clicks are likely to significantly detract from a performance and, of course, cannot be undone.

Accordingly, it is good practice not to go over 50% processor or disk load in live performance events. The optimisations in the following sections are worth considering to keep this amount of system headroom.

5.4.3.1 Optimising tracks

Modern non-real-time music productions can and often do have a great many tracks. For a live performance event, audio or MIDI tracks that will not need individual balancing or live modulation (e.g., using the systems explored in Sections 5.5 and 5.6), it may be worth bouncing/submixing/or resampling into one track or several stems. If not, each track will take up valuable processing, especially as each track is likely to have multiple plugins inserted. For tracks that need to be played and manipulated in real time, the following optimisations are recommended.

5.4.3.2 Optimising 'live' instruments

Any third party (i.e., not included with a host DAW, such as Ableton Live) instrument plugin is likely to be less processor efficient and, therefore, less reliable than a host DAW plugin. It may be worth considering whether an equivalent host DAW instrument plugin is available and acceptable. If not, does it need to be altered using its parameters, or could it be sampled and altered by a sampler (e.g., Ableton Live's Sampler or Simpler)?

5.4.3.3 Optimising effects

Similarly, for any third-party effect needed to be running in real time, it is worth checking to see whether a host DAW effect may be able to replace it.

5.4.3.4 Moving from a linear timeline into a non-linear clip launching environment

After these optimisations are explored, the clips in the linear timeline (i.e., arrangement view in Ableton Live) can be split into the size required and copied and pasted into the non-linear clip launching environment (i.e., session view in Ableton Live).

Ableton Live has a specific process for this: click-and-drag to select the section that you want to copy, then right/control click and choose 'Consolidate Time to New Scene'. A consolidated clip (i.e., just one) will be created on the same track in the session view underneath the selected slip slot.

5.4.4 Advanced non-linear launching/automation

As well as the methods in Section 5.3.2, clips can be launched in ways that allow more complex clip-launching arrangements by using virtual MIDI buses and accessing the API.

5.4.4.1 Virtual MIDI buses

A virtual MIDI bus is software that allows MIDI communication between applications. Virtual MIDI buses are handy in live performance DAWs such as

Ableton Live because MIDI notes can automate any mappable action, including clip launching.

Macintosh computers have a virtual MIDI bus built into the operating system called the inter-application communication (IAC) driver, which can be activated using the Audio MIDI setup system preference window. A free windows equivalent is the loopMIDI software created by Tobias Erichsen, which can be found at www.tobias-erichsen.de.

Virtual MIDI buses can launch and automate mappable actions using non-linear and timeline-based methods.

Non-linear methods: Figure 5.17 shows a screenshot of an Ableton Live set showing the non-linear clip launching view (session view) in MIDI mapping mode, demonstrating the use of the IAC driver to set up a cascade of automation messages. As can be seen, the first track, named '1 IAC_CONTROL', has its output set to the IAC driver. The 'CONTROL 1' clip in slot 1 contains five MIDI note on messages, shown as the black rectangles, and one CC message, shown as the blue envelope graph, in the MIDI editor view. The NTPD Max for Live plugin by ELPHNT (which can be found online at elphnt.io) is used to keep note of the mappings, and its floatpad can be seen over the mapping view. When the CONTROL 1 clip is launched, it plays a series of MIDI messages mapped to do the following: MIDI notes C-2 and C#-2 launch clips track 2 slot 1 and track 4 slot 2, respectively, MIDI note D-2 toggles record arm on track 5, and MIDI note D#-2 initially records into track 5 slot 1, then after two bars of recording

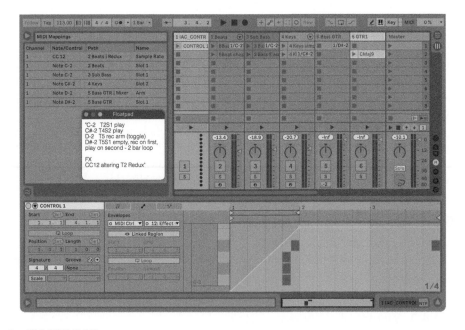

FIGURE 5.17

Ableton Live set showing the session view in MIDI mapping mode demonstrating the use of the IAC driver

another D#-2 note stops the recording and launches the clip. MIDI CC12 sweeps the sample rate of a Redux effect on track 2 from 1.07kHz to 22.7kHz over the first bar of the clip.

Timeline methods: As shown in the previous section, the non-linear clip launching method can launch a clip to cause a cascade of launching/automation. It is also possible to place the control clips (i.e., clips containing MIDI information outputting to a virtual MIDI bus) on a track running in the timeline (the arrangement view in Ableton Live). A timeline-based track can then control other tracks in non-linear clip launching mode at specified times, combining linear and non-linear approaches. The linear timeline can be used in a strictly linear fashion by starting transport, allowing the control clips to record, launch, and automate clips and parameters as they occur from beginning to end. This approach is particularly effective for automating a live looping setup (explored further in Section 5.7).

However, the timeline does not have to operate in a strictly linear fashion. Cue points or markers can be used to jump to specific times. The timeline's loop area can be manipulated to shift to different regions using plugins such as the Max for Live audio effect RB.4arrangeloops I developed (and discussed further in Section 5.5.1.3). Figure 5.18 shows a screenshot of an Ableton Live set showing the linear timeline (arrangement view) in MIDI mapping mode, demonstrating the use of the IAC driver to set up a series of timeline-based messages causing recording and playback in the session view. In Figure 5.18, the first red clip, named CTR, plays the messages C-1 on the 4 beat and D-1 on the 4&. As the mappings show, this record arms track 5, and then sets track 5 clip slot 1 recording. The green clip then plays C-1 again, which toggles record arm stopping the recording and looping the recorded clip from bar 5. The red and green clips at bars 8 and 10 do the same for track 6 clip slot 1, and the yellow clip plays D0 and D#0 which stop both tracks 5 and 6.

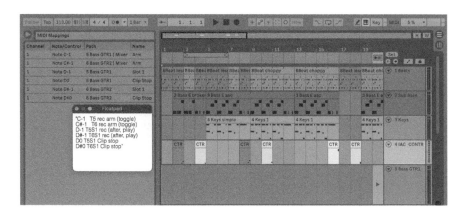

FIGURE 5.18

Ableton Live set showing the arrangement view in MIDI mapping mode demonstrating the use of the IAC driver

Figure 5.18 is a quite basic arrangement, and it is easy to see how virtual MIDI bus setups can quickly become very complicated (especially combining arrangement and session views), and it is recommended that an organised system is developed so that the behaviours required are those which occur on stage.

5.4.4.2 Accessing the API

The API allows other pieces of software to control elements of the host software in a more direct, flexible, and complex manner than a virtual MIDI bus. To give a specific example of a common Virtual MIDI bus issue in Ableton Live, mapping the record arm described in the previous section is a toggle operation. Accordingly, if the track is already record-armed, then the MIDI message will turn off the record arm, and nothing will happen when the clip is launched because it is empty.

Rather than a simple toggle operation, the API can give a specific command at a particular time – for example, my Max for Live SuperDuperLooper plugins shown in Figure 5.19 and Figure 5.20 that are currently in development. Ableton Live allows certain Max for Live objects to access the API using what they call the live object model (LOM). In Max, an object is an element of code that performs a particular task. The LOM is a list that describes the functionality and hierarchy of the objects that have access to the various aspects of the Live API.

These plugins extensively use the Live API, allowing a complex series of recordings to create live looping setups. These plugins are explored in Section 5.5.2.

The specific example in Figure 5.20 shows the part of RB.sdlooper2×3 that controls clip recording and playback. (The full patcher window is shown in Figure 4.11.) The following paragraph briefly explains how this part of the patcher works, focusing on the API elements.

The objects called 'live.object' send information to the Ableton Live API, triggering specific actions (the rectangular boxes are called objects, each doing a particular job). The left-hand 'live.object' is being sent the 'call fire' message into

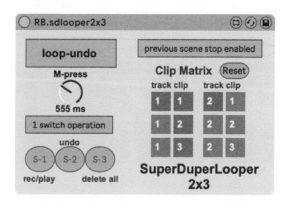

FIGURE 5.19

Ableton Live user interface of the RB.sdlooper2×3 Max for Live plugin

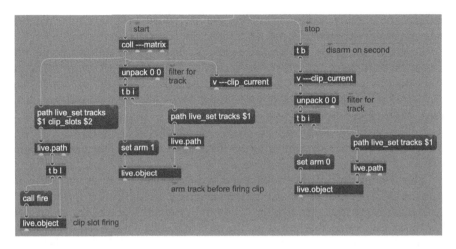

FIGURE 5.20

Lower section of max for live patcher controlling clip recording and playback

its left-hand input, which either records or plays back a clip slot (the right-hand input sets the clip slot first). The two other 'live.object's control the record arm state. The 'live.object' with the 'set arm 1' message going into its left-hand input ensures that the track is set to record arm before the clip slot is 'fired'. The 'live.object' with the 'set arm 0' message going into its left-hand input turns off the track's record arm, causing the recorded clip to playback.

As this shows, accessing the API allows for much more specific use of the DAW, so highly flexible plugins can be built. Most of my Reactive Backing plugins (discussed further in this chapter, all starting with the prefix RB. and available at www.reactivebacking.com) make extensive use of the Ableton Live API to offer a greatly expanded workflow. Many other Max for Live developed plugins are also available, several of which access the Ableton Live API. These can often be found at www.maxforlive.com.

Ableton Live and Reaper are the main DAWs currently allowing access to their API. Ableton Live uses the Max for Live programming environment, and Reaper uses a script-based system called ReaScript.

5.5 REACTIVE ELEMENTS

Reactive and chance elements are the two main categories of techniques underpinning the 'pseudo-live' musical elements described in Section 5.1.3.4 and indicated in Figure 5.2. Pseudo-live musical elements are musical parts that are not entirely live but demonstrate some significant amount of liveness as described by one or both of the practical conditions of performance liveness discussed in Section 5.1.3.1 – that the performance is:

1. unique to that moment, and
2. musically expressive.

Effective use of reactive elements makes use of performance data, which is likely to be musically expressive and is inherently unique to that moment.

Reactive pseudo-live musical elements are discussed in contrast to the non-live musical elements introduced in Section 5.1.3. One of the most basic and established arrangements to augment live performance is the use of backing tracks – either two-channel or multi-channel – running from a standalone hardware playback device or a DAW on a computer. As discussed in Section 3.5.1 (where the practical production and performance factors are discussed), the term 'backing track' refers to a fixed audio file played so musicians can play along with it.

The widespread use of backing tracks in popular music live performance events is evidence of how robust and reliable they are. In live performance, it is hard to overstate how vital system reliability is, which is one of the key reasons that there are so few alternatives to the established backing track. Another key reason, which exists alongside the need for reliability, is that establishing an alternative to an entrenched method is particularly difficult. The difficulty of establishing alternatives is mentioned in Section 4.3.2.4 in the context of the QWERTY keyboard. It comes under the economics term 'path dependence' and the psychology term 'immunity to change'.

While alternative methods are difficult to establish, they are worth developing and exploring because using a backing track dramatically restricts performance expression, especially regarding timing. The electronic alternatives to the backing track (other than playing with more musicians) can be categorised into systems that use reactive or chance techniques, discussed in this section and Section 5.6, respectively.

'React' and 'reactive': If a general model of human group interaction in performance is considered, it is clear that a group of musicians are likely to react and interact with each other in many ways. The terms 'react' and 'reactive' are used here in situations where communication is predominantly one way – for example, a musician following the tempo of a metronome. The terms 'interact' and 'interactive' are used here where communication is predominantly *more* than one way – for example, a group of musicians improvising and trading musical ideas live over a blues progression.

Many psychological and cultural factors determine how musicians react and interact. A vital psychological factor is the trait of synchronising to repeating external events (i.e., 'playing in time'), known in psychology as sensorimotor synchronisation (SMS). Incidentally, it is worth noting that SMS is not limited to music – an example is the well-known phenomenon of people walking together who tend to synchronise their walking rhythms. The cultural factors are numerous; however, an obvious example would be learned musical frameworks such as following a conductor's direction in an orchestra's framework or playing by ear in an improvisational framework of bebop jazz.

In analysing and understanding the elements that make meaningful group performances for musicians (i.e., human agents), similar arrangements can be created, allowing machine agents to react and interact with musicians. Much like in sound synthesis, however, it is likely that the more exciting uses of the reactive systems are not the recreations of existing situations but new unforeseen ones.

Following and contributing: The difference between reactive and interactive systems is demonstrated mainly by how much these systems **follow** and **contribute**.

In a human group ensemble, the amount that any musician is either leading and or following other musicians is key to the whole group dynamic. The human leading and following model can be explored and applied to an electronic music and system design context. For example, in situations where several musicians are performing live together, there is often a clear hierarchy of a band leader, musical director, or conductor who will lead the other musicians, who in turn will follow.

The most extreme example of following is where direct communication for a change of global parameter is given. Then this change is carried out in a situation where the follower cannot contribute to the leader's actions. The most obvious example is indicating changes to the arrangement – for example, using the Apito in samba music. The Apito is a high-pitched whistle that plays different patterns to indicate changes to the arrangement, which need to be followed by all musicians for the performance to continue coherently.

A different and potentially more familiar scenario, where there is no means of the following musician to contribute to the arrangement, is where the musician is playing along to a recording, a click track, or a delayed audio (and possibly video) feed. The musician in this example has to follow the click track or audio; otherwise, they will fall out of synchronisation or, in musical terms, play out of time. A fixed backing track cannot react to the musician, so the musician cannot contribute any changes to the tempo of the music as long as they are just following.

To put it another way, in both the previous arrangement variation and tempo following examples, communication is strictly one-way, so there is no chance for the system to be interactive.

In other situations between musicians where the means of communication are not restricted to being exclusively one way, the degree of following and contribution is likely to be far more nuanced. For example, when several musicians play together without a clear leader/follower hierarchy, it may be commonly agreed that one musician will be leading the timing of the music (e.g., a drummer), and the other musicians will follow that musician. Significantly, however, in this situation, the other musicians *can* contribute in less obvious ways to the tempo of the music. The drummer is still leading the timing by dictating the tempo but can react to cues from the other musicians, often instinctively, allowing the tempo to push and pull, giving the performance a far greater degree of musical expression.

It is the ability for elements to both follow *and* contribute to the music that allows interaction. The degree of following and contributing actions in live performance is likely to fluctuate in real time.

While bringing machine systems into a musical performance may initially appear to restrict this interactivity, this is not necessarily the case, as this section hopes to establish.

Global and instrument-level parameters: The musical parameters that change when musicians react and interact can be split into global and instrument-level parameters. Global parameters include tempo, dynamics, and arrangement, as these parameters similarly affect all musical parts. Instrument-level parameters include groove and part variation, as these parameters can affect musical parts differently at the instrument level.

While the general model of human group interaction in performance is the obvious place to start an exploration of reactive systems, it is clear that any parameter that can be changed is open for consideration. The other parameters that may be considered include those that belong to the sound manipulation techniques of the recording studio discussed in Chapter 2 and the electronic sound creation techniques discussed in Chapter 3. These techniques produce a very long list of parameters, but some suggestions for possible opportunities at both global and instrument levels are considered under the umbrella term of 'reactive production'.

5.5.1 Global parameters

The global parameters discussed here are the three musical parameters of tempo, dynamics, and arrangement. Additionally, the global production parameters of side chain compression are discussed as a more speculative approach.

5.5.1.1 Tempo following

Tempo is the speed of the music, generally indicated in beats per minute (bpm). For example, the tempo 120 bpm typically means that there will be 120 quarter-notes in a minute (500 ms or 0.5 seconds between each quarter-note).

Interestingly, tempo following is a particularly human trait, as noted by Repp[4]: "animals do not spontaneously move in synchrony with rhythmic auditory or visual stimuli, and there seem to have been no successful attempts to train them to do so" (p.969).

A fundamental skill for a musician is to follow the tempo of other musicians, with the metronome being a vital tool for the musician to use to practise their timing.

A tempo-following system is a system that can follow the tempo of one (or more) musicians. It is worth mentioning that score-following systems are also inherently tempo following, as they follow the notes being played from a score. Examples of these systems include IRCAM's antescofo~ (a contraction of the words 'Anticipatory Score Following'), and the educational software Tonara, both of which offer a type of tempo following.

Standard practice for playing popular music is to play by ear or from memory, not to play from a score. A musician playing popular music is typically expected to play from memory and by ear. A system of reactive backing for popular music must track the beat of the music without the reference of a score, a process often referred to as live beat tracking.

Offline beat tracking (i.e., not live, an audio file) is a standard process within music information retrieval. Live beat tracking, however, is a remarkably different process in that it needs to run in real time and can only compare to the beats in the past. A successful method of beat tracking can result in a system that can follow the tempo of music as long as the time signature and relative position of beats are known.

Ableton Live has an in-built system of tempo following called 'Follow' introduced in version 11 in 2021. This system uses an audio tracking system designed to follow percussive audio that occurs on the strong beats of the bar (i.e., the '1, 2, 3, 4'). This system works well when the audio is routed from a sound source playing loudly on the beat (i.e., a kick or snare drum). However, that sound being the driver of a system of tempo following is likely to change how that instrument is played as there are no controls to customise the tempo following.

An alternate method of controlling tempo is to use tapping data and effective customisation. I took this approach in developing the RB.tempo Max for Live plugin in 2014.[5] RB.tempo uses a process called relative tap tempo. Tap tempo uses the time delay between taps to determine a tempo but is too mechanistic to be used as a practical real-time tempo following system. The three main problems when using tap tempo to adjust the tempo are as follows:

- In the first few bars, tempo changes are likely to be very large, unmusical, and instantaneous jumps.
- In the following bars, the tempo gets progressively averaged, making the system less and less responsive as it is used.
- Every single beat must be unerringly tapped for it to work.

A relative tap tempo system takes a tap's position relative to the beat as transport is running. If the tap is ahead of the beat, the tempo is increased proportionally, and if the tap is behind, the tempo is decreased proportionally.

Some valuable insights are revealed if we can refer back to SMS, introduced at the beginning of Section 5.5. Several notable studies have found that when humans tap along with repeating tones (i.e., a tempo), the taps tend to precede tones. As Repp and Su[6] summarise, the time that the tap precedes the tone tends to be smaller for musicians than for non-musicians. Krause, Pollok, and Schnitzler[7] found that at a tempo of 75 bpm, the amount of time that the tap precedes the tone of drummers was the smallest at about 20 ms, compared to professional pianists, amateur pianists, singers, and non-musicians, which was about 50 ms.

For this reason, the RB.tempo device has an offset control to allow for taps before (or even after) the beat. A positive setting for the offset adjusts for tapping that precedes the beat.

Other customisable control parameters include a broader or more narrow control window and increased or decreased sensitivity and smoothing. A screenshot of the RB.tempo-audio plugin is shown in Figure 5.21.

5.5.1.2 Reactive dynamics

In musical instrument performance, the term 'dynamics' refers to how loudly the instrument is played. In sheet music, dynamics are indicated using performance directions ranging from *ff* (*fortissimo*, very loudly) to *pp* (*pianissimo*, very quietly). For popular music, the dynamics are typically contextual and are likely

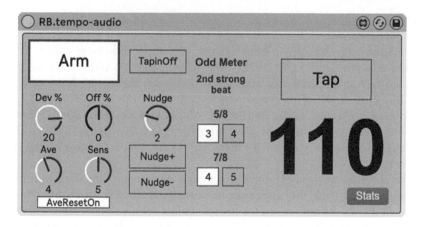

FIGURE 5.21
RB.tempo-audio Max for Live plugin

to be interpreted as relevant (e.g., ballads tend to be played quietly, and anthems tend to be played loudly).

The main difference between the sound of an instrument played loudly or quietly is the volume. However, most instruments, in particular acoustic instruments, have significantly different timbres when played loudly or quietly. For example, a snare drum is likely to have less energy in the high frequencies when hit gently and more energy in the high frequencies when hit hard. For MIDI instruments, this issue is relatively easy to address. For samplers, the use of velocity layers, as discussed in Section 3.6b.1, solves this. For synthesisers, if this sonic variation is desired, it can be achieved by offsetting the filter frequency of a low pass or high shelf filter, depending on note velocity.

The RB.dynamics plugin I developed is shown in Figure 5.22 and is designed to scale up or down the velocity of MIDI instrument notes by controlling the drive parameter of the Ableton Live MIDI effect Velocity, shown in Figure 5.23. The drive parameter offsets the velocities so that positive values increase the velocity of most notes and negative drive decreases the velocity of most notes. For audio tracks, the RB.gaindrive plugin is created to allow gain scaling equivalent to the drive setting in the Ableton Live Velocity plugin; however, this solely changes the signal level. One method that gives potentially useful timbral variation is to use an Audio Effect Rack to create a macro that adjusts the drive parameter of the RB.gaindrive plugin and the gain of a high shelf filter. This arrangement is shown in Figure 5.24.

5.5.1.3 Arrangement variation

The overall arrangement of most pieces of music, especially songs, is most likely to be determined in advance. Notable exceptions to this rule include aleatory or chance music. A good example of a minimalist piece of chance music is the Terry Riley piece 'In C', composed in 1964, consisting of 53 short melodic phrases

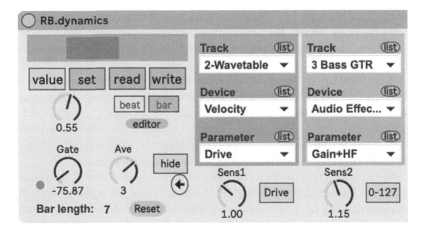

FIGURE 5.22

The settings and the first two track of the RB.dynamics Max for Live plugin showing low input level and a significant reduction in dynamics

FIGURE 5.23

Ableton Live MIDI Velocity plugin showing the drive value set negatively

FIGURE 5.24

Ableton Live Audio Effect Rack macro controlling RB.gaindrive plugin and EQ Eight plugin

played by a group of about 35 musicians, with each musician deciding how many times to play each phrase. Because of these arrangement rules, the piece is likely to be different each time it is played, and the length is not determined until the piece has finished.

In live popular music, it is common practice to alter some parts of the musical structure. The most common live variation of the musical arrangement is to extend or shorten specific sections – for example, musical parts such as breaks, solos, bridges, or the middle eight. These are musical parts between verses and choruses likely to be based on reasonably short repetitive parts (often called vamps), allowing more convenient lengthening or shortening.

Different musicians may sometimes influence the decision to change the arrangement. However, it is likely that one member of the group will be leading the decision-making to ensure coherence and reduce the possibility of confusion.

An agreed form of communication is required for the arrangement variation in a group of musicians playing freely together. For example, in the 1970 song 'Sex Machine', which was recorded live, James Brown incorporates his musical directions in the song lyrics as he theatrically asks Bobby Byrd if they should take 'em to the bridge. Another example is the apito whistle in samba music mentioned in Section 5.5.1.

The term 'variation' in this context is used to indicate a specific type of following that, unlike tempo following, is not continuous because the arrangement changes are only likely to occur at certain times and the directions to vary the arrangement are likely to be a single command per variation.

Most accompaniment systems are likely to use either linear timeline or clip-launching transport formats, so systems of arrangement variation need to do so reliably, smoothly, and conveniently.

Linear timeline variation: In typical DAWs or plugin hosts (e.g., Logic or Mainstage, respectively), a loop or repeat function (referred to here as timeline loop) is available so that a particular section of the timeline can be repeated. If only one section of music is required to be varied, then it is relatively simple to set this timeline loop in the software.

One simple and flexible approach is to create a musical phrase of the section to be looped (e.g., a simple two-bar phrase). The timeline loop mode can then be turned off, allowing the arrangement to move to the next section. It is worth noting that in some DAWs, transport is only allowed within the loop section when the loop section is activated (including Logic). In others (e.g., Ableton Live), transport can run before the loop and then remain in the loop area for as long as the loop mode is active, which is necessary if this system is to be used. A remote means of controlling the loop mode's deactivation (and activation) can be created by mapping a MIDI controller (e.g., a MIDI pad or footswitch) to the timeline loop control.

If, however, there is more than one timeline loop required (which is likely in a live performance set), then multiple timeline loops or the means to move the timeline loop conveniently during a performance are needed.

The RB.4arrangeloops plugin I developed is shown in Figure 5.25 and allows the timeline loop (called the loop brace in Ableton Live, activated and deactivated by a single loop switch) to be resized and repositioned dynamically within performance for four different loop areas: L1-4. The Ableton Live loop switch can then be freely mapped to a MIDI controller to turn the looping on and off. As many instances of the plugin as wanted can be used to allow for as many loops as required.

Clip launching variation: For DAWs that allow clip launching (see Section 4.4.2.2), musical sections are typically launched in scenes. In Ableton Live, as tracks are arranged vertically, scenes are the adjacent clips arranged in horizontal rows. Clips and scenes are most conveniently launched by a pad controller such as the Ableton Push or the Novation Launchpad, which works well for musicians whose hands are free and who work using a tabletop format (tabletop devices are discussed in Section 4.3.4.2.2). However, for musicians who want to control using their feet, it is likely that the control elements such as foot switches, pads, or pedals will need to be efficient and simplistic enough to be operated without taking up much of the musician's attention.

Because of this, a more simple and efficient layout of control elements can be created using plugins such as the RB.multiverse shown in Figure 5.26, which gives the capability to trigger different scenes with the same control element.

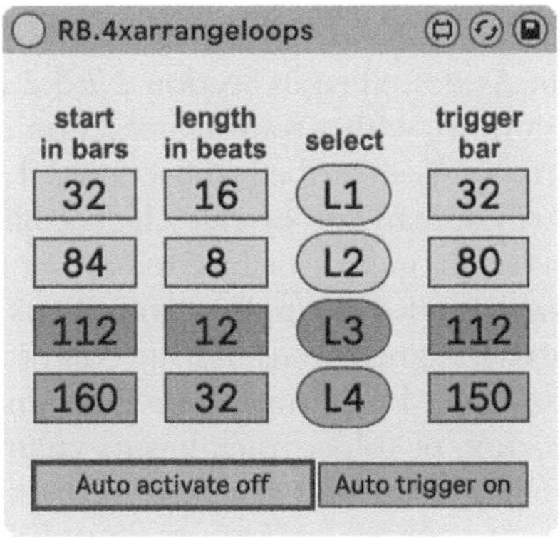

FIGURE 5.25

RB.4arrangeloops Max for Live plugin

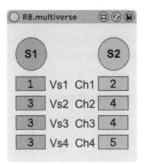

FIGURE 5.26

RB.multiverse Max for Live plugin

I created this plugin to allow two switches to control the song's arrangement using scene launching, one mapped to verses and the other to choruses (or other section types). In the example of the screenshot in Figure 5.26, the first time S1 is triggered, scene 1 is launched. The second time S1 is triggered, scene 3 is launched. The third time, scene 3 is launched again, and so on.

5.5.1.4 Global reactive production elements

Unlike the previous musical examples of live tempo, dynamic, and arrangement variation, there are no established live production equivalents to recreate.

As mentioned in Section 5.5, any parameter from the array of real-time effects can be modified based on performance data. An exhaustive list of the potential techniques and parameters for reactive elements is not likely to be helpful, as most techniques tend to be specific to a discrete sound manipulation process but may nevertheless be worthy of experimentation. Side chain compression (introduced in the context of recording studio techniques in Section 3.1.2.4) is focused on here as an example of a suitable global reactive production element. Side chain compression is chosen mainly due to the increasing familiarity of the technique in genres beyond popular electronic music and because it is well suited to be applied to a combination of signals.

Side chain compression: As described in Section 2.2.3.2.2, compression reduces peak signals by a certain ratio, with a transient reaction determined by the compressor design and settings. As described in Section 3.1.2.4, side chaining uses an alternate input. A widespread use of side chain compression from popular electronic music is to have a compressor inserted into a track that can manipulate a combination of signals – for example, a group, aux, or folder track (or the master bus for particularly aggressive use). This compressor then has its input side chained to the kick drum. Every time there is a signal from the kick drum, any signal in the group, aux, or folder track has its volume reduced briefly. The result is a rhythmic swelling effect that accentuates the kick drum (or any other side chain signal).

In the recording studio, the amount of this compression will be finessed to achieve the desired effect; however, in a live situation, the musical elements are not likely to be anywhere near as controlled. Despite that, parameters of the side chain compressor, such as threshold, could be controlled by performance parameters such as the dynamic levels using the RB.dynamics plugin discussed in Section 5.5.1.1.2. Over a certain dynamic, the side chain compression would become apparent and give the performance an additional edge.

5.5.2 Instrument-level parameters

Instrument-level parameters are instrument-specific parameters that vary individually from musician to musician within a group. This variety of individual expression is one of the main factors that give a musical group a complex dynamic, especially in a live performance environment. These parameters can be summarised as groove and part variation.

5.5.2.1 Groove

As well as the global musical parameters of tempo and dynamic (as discussed in Section 5.5.1), a musician is likely to impart a certain groove. In specific musical terms, 'groove' refers to timing variations away from regular time divisions within musical phrases. For example, these divisions may be 1/16th notes and are reflected in the grid and quantise settings in a DAW.

Typically, the simplest and shortest of these phrases is the musical bar. A musician may have a particular groove where they play slightly ahead of the beat at one part of the bar (e.g., on the two), and then slightly behind the beat at another part of the bar (e.g., on the four-and). While a musician may tend to play grooves based on their own style of playing, this is also musical piece/song, genre and instrument dependent.

DAWs allow for the capture and application of grooves in a manner generally best suited to offline editing. Figure 5.27 shows the Ableton Live groove pool containing two grooves. Base is the grid used to set timing resolution and Quantize sets how much the groove will be set to the grid. The other controls indicated the groove can vary the parameters of an audio or MIDI clip. Timing sets how much the groove's timing will change the clip it is applied to, Random sets how much timing fluctuation is applied, and velocity sets how much the groove velocity will change the clip velocities. A groove is applied by dragging and dropping it onto the clip. The settings of the groove can be adjusted in real-time non-destructively or 'committed' to the clip, which 'writes' the variations to the clip.

Grooves are well suited to an offline editing workflow, allowing timing tweaks as parts develop. However, it is a rather blunt tool for applying grooves in a live instance. The groove will ideally need to be set up in advance, and while the Timing % can be mapped to performance data that indicates either more or less of a groove, this is likely to act more like a chance system than a convincingly natural groove (chance systems are discussed in Section 5.6). If the clips to be varied using grooves are being recorded live, this adds another layer of difficulty because drag and dropping is a good example of a typical offline editing action rather than an action well suited to live performance.

A potential fix to the problem of bringing a more live groove to pseudo-live elements would be to train a machine learning process to analyse musicians'

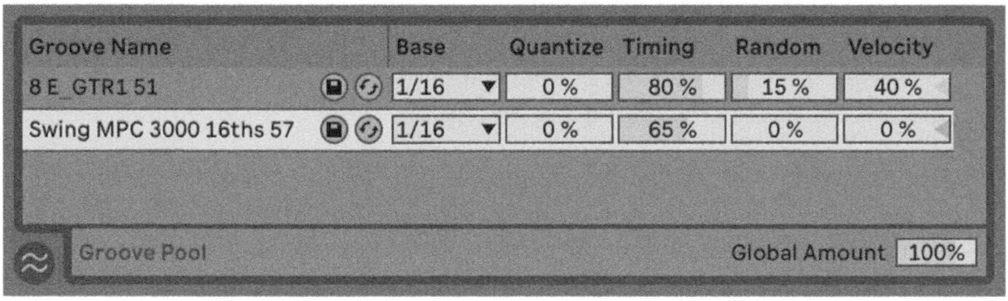

Groove Name		Base		Quantize	Timing	Random	Velocity
8 E GTR1 51	🔘 🔄	1/16	▼	0 %	80 %	15 %	40 %
Swing MPC 3000 16ths 57	🔘 🔄	1/16	▼	0 %	65 %	0 %	0 %
〰 Groove Pool						Global Amount	100%

FIGURE 5.27

The Groove Pool in Ableton Live

timing and then apply that data to generate an evolving groove to machine-driven musical elements. Currently, no such systems are widely available.

5.5.2.2 Part variation

Most musical parts that are not improvised are likely to be written, modified, and refined before a performance. When performed, they may be played with a degree of variation beyond the micro timing variation of groove – that is, small changes to the notes and hits themselves. These changes are likely to be subtle – for example, adding some passing notes as embellishments or swapping specific notes in a passage. These variations will be dependent on genre, aesthetic goal, and personal style.

If a linear timeline system is used for accompaniment, part variations are likely to be programmed beforehand, leading to an approach likely to result in remarkably static accompaniment parts.

An approach that does allow for specific variations is to use a clip-launching environment and create several different clips that are variations of the musical parts. These clip variations can be triggered using chance settings, performance data, or a combination. Part variation is an example of a situation well suited to chance variation. The specific chance elements relating to part variation are discussed further in Section 5.6.2.

Performance data can be used to replicate human characteristics of part variation in performance, giving some simple rules or weighting to the chance settings to more closely replicate how a musician is likely to vary their parts. For example, it may be observed in a specific group of musicians that variations are more likely to occur later in the music after the

identity of the musical parts has been established, so the chance for these variations to happen could be increased. It may also be observed that parts are embellished over or under specific tempos or dynamics.

While no devices exist at the time of writing that have this ability, it is one of several additional research projects for my Reactive Backing work.

5.5.2.3 Instrument-level reactive production elements

Similar to the global reactive production elements discussed in Section 5.5.1.4, there are no well-established live production equivalents to recreate at the instrument level. As also discussed in Section 5.5.1.4, an exhaustive list of the potential techniques and parameters for reactive elements is unlikely to be helpful. Instead, the method of side chain gating is focused on here as an example of a suitable instrument-level reactive production element.

Side chain gating: As also described in Section 2.2.3.2.2, gating removes unwanted sound in a signal – specifically the noise between louder signal elements. However, if a gate is side-chained like a compressor, as described in Section 5.5.1.4, the gate can be opened and closed by another signal level. Gates open when the input signal is above the threshold, but some gates also allow their behaviour to be reversed so that signals pass when the level is *below* the threshold (the Gate in Ableton Live uses the 'Flip' control to achieve this).

The result is a far more aggressive cutting off of the signal than side chain compression, which typically sounds like a particularly choppy rhythmic effect. There is also the possibility of being side chained to a live signal, bringing more liveness to the part and acting much like the extensive creative techniques discussed in Section 5.2.3.

5.5.3 Interactivity

The reactive systems described in the previous sections may seem to be *exclusively* reactive situations. However, how the system reacts and the characteristics of the music played by the system (i.e., how percussive, textural and melodic it is) *are* likely to contribute somewhat to the nature of the performance. For example, a tempo-following system that smooths out tempo changes can influence other musicians to follow these changes more slowly. Alternatively, if there is little smoothing and high tempo change sensitivity, musicians are likely to be more alert to faster variation in the tempo. Accordingly, a reactive system is likely to contribute in some way to the performance, calling into question a category that is exclusively 'reactive' – that is, that it only follows and does not contribute. How these systems could contribute as well as follow, and how they may develop into systems that can significantly interact, are areas worthy of consideration and further research.

5.5.4 Reactive DAWs

5.5.4.1 Tempo following in a DAW

For a DAW to incorporate a system of tempo following, it needs to be able to adjust the timing of the MIDI and audio. Different DAWs manage timing differently, but smooth MIDI and audio data adjustment are essential to using any DAW as a reactive system.

Because they are simply a message triggering an action, changing the timing of MIDI notes (or other MIDI data) is a relatively simple operation. Recorded audio, however, is fixed in time. Changing the start time or stopping the audio file from playing is easy; however, adjusting the length of an audio file requires the process of time stretching (discussed in Section 3.3.1.5.2 and also referred to as warping in Ableton Live), which is a far more complicated process.

Some DAWs are designed to offer audio and MIDI manipulation optimised for live performance, such as Ableton Live, FL Studio and Bitwig. Time stretching (or warping) does have limits, however. The algorithm used needs to be matched to the audio being stretched – the options typically being algorithms that are optimised for either percussive, pitched monophonic, or more complex audio. Even when matched to the most relevant algorithms, if audio is shortened or overextended, digital artefacts will appear more pronounced the more the audio is stretched.

5.5.4.2 Specific implementation of reactive systems in a DAW

It is common to map MIDI controller keys, pads, dials, and faders to a wide range of parameters in most DAWs. However, for reactive systems, an additional program or device is needed to map a more complicated input (e.g., instrument signal level) to the output (e.g., the dynamic level of a synthesiser patch).

One method, as used by the Reactive Backing RB. plugins, is to access the DAW's API. (Accessing the Ableton Live API using Max for Live is explored in Section 5.4.4.2.)

5.6 CHANCE ELEMENTS

As mentioned at the beginning of Section 5.5, as well as reactive elements, chance elements are the other main category of techniques that can be used to create pseudo-live musical elements. Chance elements can incorporate unique variations to music, satisfying the first condition of performance liveness (see section 5.1.3.1): that the performance is unique to that moment. The second condition of musical expression is absent in chance systems (that are not reactive). However, there is always a degree of random variation in human musical expression.

The act of musical expression can be seen as intentions managed by the embodied mind that are sent through the nervous system to muscles and tendons that carry out these intentions as closely as possible. Considering this complex arrangement, it is clear that there will always be an amount of error, noise, or what could be described as 'micro-chaotic variation' in human liveness. While deliberate practice may reduce this random variation, this complex dialogue of managing the micro-chaotic variation between mind, body, and instrument is likely to be a large part of what makes an excellent performance and a critical factor in human liveness. Considering this, it makes sense to replicate this variation using chance systems to afford a degree of machine liveness.

This section discusses techniques for machine liveness using chance systems to vary pre-prepared and fixed musical elements. These musical elements are typically collections of notes and percussive hits. (For brevity, both notes and percussive hits are referred to as just 'notes' from this point onwards.)

Chance systems for varying pre-prepared musical elements can be separated into three levels:

- the grouping of notes into phrases (or clips),
- the notes themselves, and
- the specific parameters of individual notes (e.g., the timing and velocity of each note).

In this context, a 'musical phrase' means anything more than one note or hit on one track, typically the contents of a MIDI or audio clip.

As previously mentioned, MIDI clips offer a far broader and more convenient range of manipulation possibilities than audio clips, as the MIDI note message is just the instruction to play the note. Because of this, in Sections 5.6.2 and 5.6.3, all of the sections are discussed in terms of MIDI techniques. In Section 5.6.1, clips are discussed in terms of MIDI or audio techniques.

5.6.1 Phrases and chance

The ability to set up a system of launching a clip based on chance is a significant benefit of a clip-launching DAW. Ableton Live uses 'follow actions' that allow a series of clip-launching choices as one clip follows another.

Figure 5.28 shows the follow actions for the MIDI clip 'Sparse Beats'. As shown, two follow actions (under the Follow Action button) can be set, and each is given a percentage chance of occurrence. For example, in Figure 5.33, follow action A is 60%, so follow action B is 40% (although this is hidden under the list of possible actions).

This way, different MIDI or audio clips containing musical phrases can be triggered in a series of cascading randomness from simple to highly complicated. As well as the clips and their follow actions, the clips can contain notes with different chance settings.

5.6.2 Notes and chance

There are several systems for adding machine chance variation at the musical note level. As mentioned in Section 5, these systems tend to be exclusively MIDI based because of the flexibility of MIDI messages. Note-level systems can be separated into three groups: playing probability, note variation, and note generation.

5.6.2.1 Playing probability

A system of playing probability is one where the MIDI notes are already inputted; however, there is a probability given for each note to determine how likely it is that note will be played. For example, in Ableton Live, underneath the velocity lane is an additional chance lane where a percentage probability can be given to

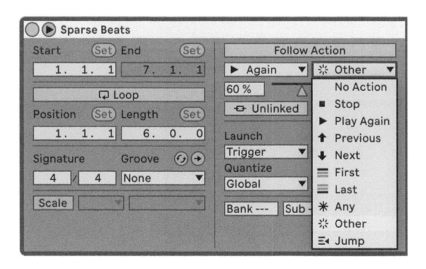

FIGURE 5.28
Ableton Live follow actions for the MIDI clip 'Sparse Beats'

each note. These percentages can be seen in Figure 5.29, where the hi-hats on the 'and' are set to between 50% and 80%. The hi-hat on the '1 &' is set to 50% chance, so there is a 50/50 chance (one out of every two on average) that it will play every time the clip repeats. The hi-hat on the '4 &' is set to 80%, giving it an 80/20 chance (an average of four out of every five).

5.6.2.2 Note variation

A system of note variation uses MIDI notes that are either played live or in MIDI clips; however, these notes are processed by a system of chance. The Ableton Live MIDI effect Random, shown in Figure 5.30, allows real-time pitch variation of MIDI notes in semitones, with the Scale and Choices controls setting the step size and range of randomness.

The Ableton Live MIDI effect Arpeggiator, shown in Figure 5.31, separates the MIDI notes chords mechanistically. Arpeggiator also offers a potentially more constrained form of randomisation if the style is set to one of the random settings.

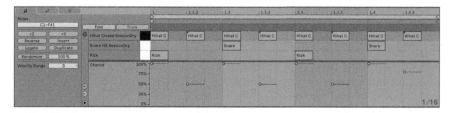

FIGURE 5.29
Ableton Live clip probability for each note (hi-hat)

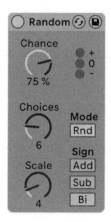

FIGURE 5.30
Ableton Live Random MIDI plugin

FIGURE 5.31
Ableton Live Arpeggiator MIDI plugin

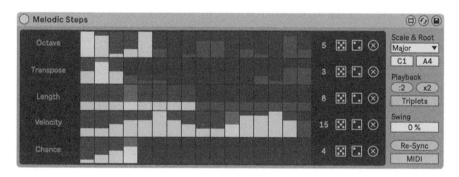

FIGURE 5.32
Ableton Live Melodic Steps MIDI effect

5.6.2.3 Note generation

In a system of note generation, MIDI notes are created by the system itself. Many systems are capable of this, each using various techniques to create a different series of notes. The standalone application Wotja by Intermorphic is particularly noteworthy as what they call an 'All-in-one Generative Music' Player & Composer Lab' and is particularly adept at creating ever-changing ambient music.

The Ableton Live MIDI effect Melodic steps shown in Figure 5.32 uses a series of alternating steps to create MIDI notes. Each step sets a parameter of a created MIDI notes, and if there are different numbers of steps for each one, they quickly shift out of phase and create different patterns, which are shown by the steps coloured blue.

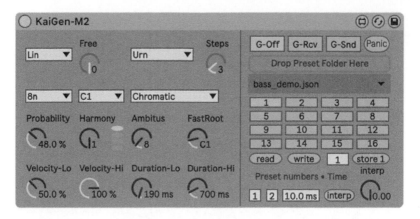

FIGURE 5.33
KaiGen-M2 Max for Live plugin

The KaiGen plugins created by musician and academic Jeff Kaiser are an excellent example of generative randomised plugins and are available for whatever price the buyer wishes to pay. The KaiGen-M2 plugin shown in Figure 5.33 generates MIDI notes dependent on various rhythmic random types and scale settings.

5.6.3 Note parameters and chance

Methods of using chance to vary the individual pre-prepared musical parameters of notes can be separated into three groups: timing, velocity, and other parameters.

5.6.3.1 Timing

In most DAWs, the rhythmic patterns of musical events are typically referred to as grooves. A straight 1/8ths groove would be eight 1/8th notes occurring precisely on the 1& 2& 3& 4& timings. An interesting rhythmic pattern that pushes and pulls in some specific way can be captured and saved as a groove and applied to other musical parts, giving it a degree of that rhythm, feel, or groove. As a part of this rhythmic management, the timing of the groove can also be randomised. Ableton Live uses a % randomised setting for its grooves (discussed in Section 5.6.2.1), which can be seen in Figure 5.30. It is worth noting that for this random function to work in a live fashion in Ableton Live (i.e., to be different for every different playback), the groove needs to be set, and the commit function should not be used. The commit function will write one instance of the random settings into the MIDI clip, and the groove will be removed from the clip settings.

If the randomisation applied to the timing is set a little too high, the timing of the part will sound like errors in playing. If set far too high, the timing will lose rhythmic meaning altogether. An appropriately small amount of randomisation is likely to give a certain kind of consistent micro-variation dependent on the musical context. The most convenient way to provide variation to the amount of variation would be to automate this setting; unfortunately, at the time of writing (version 11), Ableton Live cannot automate the random % setting of grooves. However, the random % for grooves is accessible by the Live API, so this could be built into a Max for Live plugin and is another potential avenue of research and development.

5.6.3.2 Velocity

As velocity is a fundamental parameter of the MIDI note message format (see Section 4.2), altering the velocity of notes is a relatively straightforward matter and well suited to chance systems. In Ableton Live, these methods are separated into:

- global by using the Velocity MIDI effect, and
- individual by setting a velocity range for each note.

5.6.3.2.1 Global (Velocity MIDI effect)

Inserting the Velocity MIDI effect plugin on a MIDI track allows real-time randomisation of each note's velocity as they are played through a track. Unless the random setting is automated or modulated to vary over time, the range of random variation will be the same for each note.

Figure 5.34 shows the Ableton Live MIDI effect Velocity, with a setting of 30%. This 30% random velocity variation is shown in the transfer curve as a light grey area indicating where potential velocities may occur.

5.6.3.2.2 Individual, clip-based velocity range

Individual ranges for velocity randomisation can be set using the Velocity range setting in the notes tab. Figure 5.35 shows a screenshot of a MIDI clip with a simple rock beat. The different ranges can be seen as the shaded areas under each note's velocity. (Velocity range number box shows * because multiple different velocity ranges are selected.)

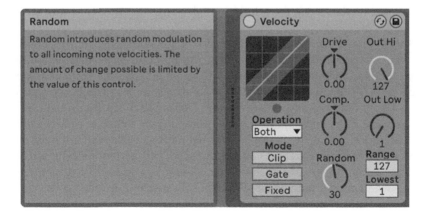

FIGURE 5.34

Ableton Live Velocity MIDI effect showing the info view of the random control

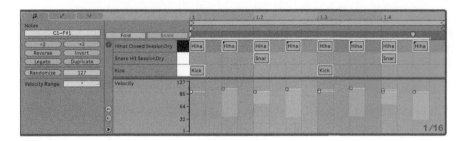

FIGURE 5.35

Ableton Live MIDI clip showing multiple velocity range settings

Using a velocity range allows an individual velocity range for each note, unlike the Velocity MIDI effect, which, even if it were automated for each musical event over time, would still apply the same amount of randomisation for simultaneous musical events. So, if some elements need a consistently high velocity on the 1, this could be given no velocity range (as shown in the Kick in Figure 5.35) and would not be randomised. Other notes, however, can demonstrate machine liveness using a velocity range. The example in Figure 5.35 shows the velocity ranges of the hi-hats on the & with a wide range of -70 and the snare hits with a smaller range of -30.

5.6.3.3 Other note parameters

As discussed previously, the timing randomisation of groove settings and the velocity randomisation of MIDI clips have specific built-in methods of modulation within Ableton Live (shown in Figures 5.27 and 5.32, respectively). Other note parameters can be modulated by mapping them by the stock Max for Live audio plugin LFO. The LFO effect is, as the name suggests, a continuous modulation source that can be mapped to an Ableton Live or third-party plugin parameter (the destination). The waveform of the LFO has a random setting, which can be randomised further using the relatively high-frequency jitter function, as shown in Figure 5.36.

Destination parameters for the LFO plugin are mapped by clicking on the 'Map' button and then clicking on the destination parameter (shown in Figure 5.33 as mapped to 'Detune'). Most plugin parameters are mappable, which gives a potentially huge list. Immediate examples include the volume or pitch of a sample (as mapped in Figure 5.10). MIDI 1.0 pitch bend and CC messages are more difficult to modify using chance systems and mapping, as the expectation is that these will be external messages that will not benefit from chance modification. External control messages can be modified by setting the reactive backing plugin RB.CCmodfilter, shown in Figure 5.11, to the relevant CC message and mapping an LFO to either the amount or offset control (for an additive or scaling modification, respectively).

FIGURE 5.36

Ableton Max for Live LFO plugin showing both a random waveform and an amount of jitter

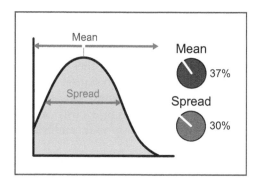

FIGURE 5.37

Design for a user interface of a random number generator

5.6.3.4 Chance control

A method of randomisation that offers more control is to use a distribution curve that allows the weighting of randomness to be adjusted. An example of a user interface for a random signal generator distribution curve applied is shown in Figure 5.37, based on the random number generators of The Mangle – a sadly no longer supported granular synthesiser VST plugin. If the spread is set to 100%, the random signal values will be evenly weighted and completely random, and the display will show a flat horizontal line (rather than a curve). With a spread set less than 100%, the mean setting pushes the weighting towards low values for settings less than 50% and high values for settings higher than 50%. The settings in figure show a fairly narrow spread (30%) weighted to somewhat lower values (38%).

As the use of randomisation becomes more widespread, it seems likely and desirable that more sophisticated systems such as this will be used.

5.7 LIVE LOOPING

A loop is a repeating section of audio, typically a phrase sample (rather than a one-shot sample as discussed in Section 3.6b) or an audio clip. Section 5.4 discusses the typical clip launching methods where the clips are often loops.

The term 'live looping' in this context is the process of capturing elements of a live performance and looping them.

5.7.1 Loop stations

The simplest and most reliable method of live looping is to use a single standalone loop station – either a pedal or a tabletop device, as discussed in Section 4.3.4.3. The audio captured by a typical loop station will typically be fixed and repeated, forming a locked timebase. This method can be particularly effective for creating a solid and repetitive backing that can be layered as the performance progresses. As with a fixed backing track, fixed live loops are non-live as they are no longer unique to that moment in time, repeating the same musical expression. This consistently repetitive and tempo-consistent form of accompaniment is not necessarily undesirable, more an aesthetic choice.

One of the critical features of a loop station is how many different tracks it allows. The number of tracks sets how many other loops it can capture and playback at once. For example, a one-track loop station can typically capture audio and overdub it. In the context of loop stations, 'overdubbing' means combining incoming audio with the audio loop that is previously captured and playing. This process can be repeated, rewriting the loop so it can be layered into a very complex audio section.

Loop stations with multiple tracks allow the creation of more than one loop. These separate loops can be recorded, overdubbed, played back, or stopped as required, giving more flexibility and, accordingly, more complexity in operation.

As a default setting, it is common for multiple track loop stations to have the loop length set by the first loop created. Consequently, if the music is played in time, the loops will stay locked in sync, starting and finishing simultaneously.

Some multiple-track loop stations allow tracks to be set to different modes of operation so that loops of different lengths can be created. For a loop station to create loops of different lengths, it needs to be set up with the required tempo and the bar length of each loop on each corresponding track. Each of these loops running on each track may also be set to synchronise with the loop station's tempo (and phase) or to be free running. Additionally, the tempo of the loops can be varied – typically by using a tap tempo function or manual input, or by connecting the loop station to another host using MIDI clock sync (see Section 4.3.3.5 for a description of MIDI clock sync and the terminology involved).

Multiple loop stations (and other audio systems) can be connected using MIDI clock sync, with the host loop station controlling transport and tempo.

Loop stations are likely to afford the human musical-visual agency of a stomp box, although not as much as a noticeable change to heavy distortion, for example. On the other hand, live looping has become a recognisable performance technique, especially among singer-songwriters – a good example of a modern musical-cultural signified. Loop stations then rank below the 'dramatic switch to a heavy distortion' scenario and above the 'switch to one or more subtle variations of tone' scenario in the general equipment hierarchy and human musical-visual agency of Section 5.1.4.

5.7.2 Live looping with DAWs

A live looping setup driven by a DAW offers a roughly opposite set of affordances compared to a setup driven by a loop station. A loop station affords an immediate and intuitive looping workflow managed by relatively few controls but is proportionally simple. In contrast, a DAW is a far more complicated setup using multiple pieces of equipment but can offer a massive range of performance opportunities. Specifically, in a DAW, loops can be reactive and chance based using the techniques explored in Sections 5.5 and 5.6.

5.7.2.1 Looper

Ableton Live can operate like a loop station if one or more tracks in Ableton Live have the Ableton Live Looper plugin inserted and the 'Multi-Purpose Transport Button' is mapped to a physical trigger (e.g., a key, pad, or footswitch). Ableton Live then becomes a highly customisable loop station set up with all the extended playback functionality of a DAW. While less procedural than the workflow of a standalone loop station, this setup still has limitations inherent in the loop station paradigm. These limitations include only one loop per track (unless the loop is dragged into a clip slot), and if the Looper controlling tempo/transport needs to be changed, this is relatively difficult.

5.7.2.2 Clip based looping

A more flexible and less procedural approach is to make use of a virtual MIDI bus to create loops in clip slots. Using the method described in Section 5.4.4.1, MIDI notes can be set up to trigger the recording and playback (and stopping) of loops into clip slots as required. These MIDI notes can be placed in arrangement (or even session) clips, allowing a highly customisable form of loop automation. The MIDI notes can also be coming from external controllers; however, this tends to get complicated quickly and can be quite difficult to manage in a live environment. Using the clip slots gives the full functionality of the DAW, including the reactive and chance elements discussed in Sections 5.5 and 5.6.

Another approach with even less procedurality is to use one or more instances of the Reactive Backing Super Duper Looper plugin (the RB.sdlooper). This plugin allows a single mapped physical trigger to record loops into a grid of clip slots. Figure 5.19 in Section 5.4.4.2 shows a screenshot of the RB.sdlooper2×3 plugin. This grid allows three clip slots on two tracks for sequential live looping. Different grids of various dimensions allow other looping arrangements to expand the loop station/live looping paradigm.

Both approaches of the Looper and RB.sdlooper plugins (and mapping MIDI notes to record loops in clips slots) are likely to be operated by a MIDI controller, which is often floor-mounted. As discussed in Section 5.1, the flexibility of MIDI controllers may indicate a reduction in human musical-visual agency. However, live looping has become a more mainstream technique thanks to its extensive use by musicians such as Ed Sheeran. Much like the causally overt distortion effect applied with the press of a footswitch, live looping can also be performed in a more causally explicit manner.

5.8 CODING AS LIVE PERFORMANCE

The final area for consideration of music technology in live performance is live coding, which often takes place at events called algoraves. Live coding is when musician programmers create music by creating text-based computer code in a live environment. Typically, at algoraves, the screen showing the code (often a laptop screen) is projected for the audience to see as the code is created. While the music is not being performed in the traditional sense, the performance is the live programming, making the audio (and possibly video).

In this context, there is a high amount of direct causality, and the performance signifiers are literally writ large as code. However, interpreting these disembodied performance signifiers requires a highly specialised awareness of the coding language used (e.g., Super Collider, ChunK, or Tidal), and the ability to keep track of the musical context as the coding develops.

Algoraves are a highly specialised example of the electronic music performance mystique, illustrating the human tendency to infer narrative. We may not grasp what we see; however, as described by Daniel Kahneman,[8] physiological studies have repeatedly revealed that humans tend to be very quick to create causal relationships, with little regard for fact or fiction. (These ideas are explored with a more theoretical framework in my forthcoming academic book on liveness in modern music performance.)

CONCLUSION

This book attempts to establish a complex set of best practices and explore experimental approaches to augment live performance events. In doing so, there are four underpinning themes:

- There is no one set of techniques or tools that *has* to be used in any creative work. Rigid and arbitrary 'should' factors are likely to obstruct effective work.
- Digital tools are as valid as their physical counterparts for creating meaningful live performances. A violin bow, a piano key mechanism, and a DAW all offer different kinds of agency. Technological agency is unavoidable in modern music and is just as relevant as any other cultural influence. Critical involvement of technological agency is far more beneficial than arbitrarily elevating physical or analogue work with no machine agency.

- Categorising and separating concepts is helpful for understanding concepts and techniques; however, the skilful interweaving of different concepts and techniques is likely to create consistently innovative work.
- Changing set behaviours is likely to be challenging. Building creative confidence and maintaining a flexible and innovative mindset and workflow take work. However, without change, creative work becomes dull. This book attempts to empower the reader to harness the powers of technology push (stuff to do new things with) and creative pull (ideas to create new forms of expression).

It is hoped that this work forms a valuable resource, and that the ideas within it might help to spur on more innovative creative work – and maybe even result in more exchange of energy.

Let me know how you get on.
Tim

ENDNOTES

1. Kasparov, G. (2017). *Conversations with Tyler Podcast. Garry Kasparov on AI, Chess, and the Future of Creativity EP. 22* [Interview]. https://conversationswithtyler.com/episodes/garry-kasparov/
2. Exile, T. (2015). Tim Exile keynote and performance at Innovation in Music Conference. https://www.youtube.com/watch?v=jp_60_B7pNQ
3. Repp, B. H. (2005). Sensorimotor synchronization: A review of the tapping literature. *Psychonomic Bulletin & Review*, 12(6), 969–992. https://doi.org/10.3758/BF03206433
4. Repp, B. H. (2005). Sensorimotor synchronization: A review of the tapping literature. *Psychonomic Bulletin & Review*, 12(6), 969–992. https://doi.org/10.3758/BF03206433
5. Canfer, T. (2016). A System of Reactive Backing for Live Popular Music. In R. Hepworth-Sawer, J. Hodgson, P. Justin L., & T. Robert (eds.), *KES Transactions on Innovation in Music* (pp. 26–36). Future Technology Press.
6. Repp, B. H., & Su, Y.-H. (2013). Sensorimotor synchronization: A review of recent research (2006–2012). *Psychonomic Bulletin & Review*, 20(3), 403. https://doi.org/10.3758/s13423-012-0371-2
7. Krause, V., Pollok, B., & Schnitzler, A. (2010). Perception in action: The impact of sensory information on sensorimotor synchronization in musicians and non-musicians. *Acta Psychologica*, 133(1), 28–37. https://doi.org/10.1016/j.actpsy.2009.08.003
8. Kahneman, D. (2012). *Thinking, Fast and Slow*. Penguin.

INDEX

Note: Page numbers in *italics* refer to figures; page numbers in **bold** refer to tables.

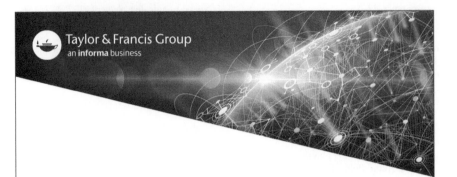